Science 3–13

Which factors have been influential in developing science teaching and learning for the 3–13 age group in the last twenty years? How might these factors have an impact on the future direction of science teaching and learning for this age range into the twenty-first century? How can teachers cope with the changes?

Science 3–13 explores some of the historical antecedents of the current position of science in the lives of younger children. It covers the various influences, from both within and outside the teaching profession, that have shaped the current science curriculum. Current practice is examined and, on this basis, speculations are made about the future position and direction of this important subject.

The contributors each cover a particular aspect of science for the 3–13 age range but common themes emerge, such as the influence of government intentions, particularly through the development of the National Curriculum. The role of research groups and the impact of IT on the development of science in schools are also examined. The views of the teaching profession as to what should be taught and how science and science teaching should be viewed within society are shown to be important factors in the mix that contributes to change.

This book forms part of a series of key texts which focus on a range of topics related to primary education and schooling. Each book in the **Primary Directions Series** will aim to review the past, analyse current issues, suggest coping strategies for practitioners and speculate on the future.

Paul Warwick is a Lecturer in Primary Science at Homerton College, Cambridge. He previously taught in primary schools and was an Advisory Teacher for Science. His research interests are IT in science, children's understanding of the procedural aspects of science, and the assessment of children's learning.

Rachel Sparks Linfield is a reception teacher at St John's College School, Cambridge, and was previously a Lecturer in Primary Science at Homerton College, Cambridge. She has written widely in the area of early years education, and her research interests include assessment of early years science.

Primary Directions Series
Series Editors: Colin Conner, School of Education, University of Cambridge, and
Geoff Southworth, Department of Education Studies and Management, University of Reading

Assessment in Action in the Primary School
Edited by Colin Conner

Primary Education – At a Hinge of History?
Colin Richards

Science 3–13
The Past, the Present and Possible Futures
Edited by Paul Warwick and Rachel Sparks Linfield

Science 3–13

The Past, the Present and Possible Futures

Edited by Paul Warwick and Rachel Sparks Linfield

London and New York

First published 2000
by RoutledgeFalmer
11 New Fetter Lane, London EC4P 4EE

Simultaneously published in the USA and Canada
by RoutledgeFalmer
Garland Inc, 19 Union Square West, New York, NY 10003

RoutledgeFalmer is an imprint of the Taylor & Francis Group

Typeset in Times by Taylor & Francis Books Ltd
Printed and bound in Great Britain by
TJ International Ltd, Padstow, Cornwall

British Library Cataloguing in Publication Data
A catalogue record for this book is available from the British Library

Library of Congress Cataloging in Publication Data
Science 3–13 : the past, the present, and possible futures / [edited by] Paul Warwick and Rachel Sparks Linfield
p. cm. – (Primary directions series)
Includes bibliographical references and index.
1. Science–Study and teaching (Elementary)–Great Britain.
2. Science–Study and teaching (Secondary)–Great Britain.
I. Warwick, Paul, 1956–. II. Linfield, Rachel Sparks. III. Series.
LB1585.5.G7 S33 2000
500'.71'2041–dc21 99–049509

ISBN 0–415–22786–0 (hbk)
ISBN 0–415–22787–9 (pbk)

Contents

Illustrations

Figures

Tables

Contributors

Penny Coltman Penny was a secondary science teacher before discovering through her own family that working with young children was her first love. She has since accumulated wide experience throughout the range of early years education as a teacher and freelance consultant. She is now a Lecturer in Science and Mathematics Education at Homerton College Cambridge, specialising in the early years. She was part of a small team of writers who produced the Longmans *Science Connections* scheme and is a regular contributor to *Child Education* magazine.

John Hobden John taught in a variety of primary schools for over twenty years before joining Homerton College Cambridge to spend more than two years co-ordinating the Primary Teacher as Scientist (PTAS) project. He has been involved in the professional development of teachers for many years, and the PTAS project has provided the opportunity to work with teachers located throughout Britain, each of whom has undertaken their own scientific research. John has published training materials for teachers, articles related to pupils' perceptions of science, and a book on railway history. He has a particular interest in the use of ICT in education and is currently an advisory teacher working for the SchoolNet 2000 Project.

Angela McFarlane Angela is Director of the Centre for Research in Educational ICT (CreICT) at Homerton College Cambridge. Her own research interests include the use of IT to enhance learning. She has an international reputation in the field of ICT in education. She led the team which developed three successful science education software applications, and has jointly published papers which examine their impact on pupils' learning. She was one of a handful of British science educators who were invited to contribute to *Kluwer International Handbook of Science Education*, published in 1998. She is the editor of *Information Technology and Authentic Learning*, published by Routledge in July 1997. In 1998 she was elected to a Readership in Educational ICT.

Ian Morrison Ian was a secondary science and physics teacher before joining Homerton College Cambridge as a Lecturer in Science Education. He has lectured in physics and the teaching of science to students in initial teacher training and has been involved in a wide range of in-service provision, including training for secondary school heads of science, training in physics for secondary school teachers and training in the physics knowledge aspects of science with primary school teachers on numerous Advanced Diploma courses. His research into children's understanding of science ideas fitted into the framework of professional development courses promoted by the DfEE.

Michael Reiss Michael taught in schools until 1988 and then became a lecturer, first at the University of Cambridge Department of Education and then at Homerton College Cambridge, where he is currently a Reader in Education and Bioethics. His research interests are in the fields of science education, sex education and bioethics, and he has directed a number of research and curriculum projects in science education. Among his books are *Science Education for a Pluralist Society* (Open University Press), *Improving Nature? The Science and Ethics of Genetic Engineering* (Cambridge University Press) and *Ecology: Principles and Applications* (Cambridge University Press).

Rachel Sparks Linfield Rachel is a reception teacher at St John's College School, Cambridge. Prior to this she was a Lecturer in Primary Science at Homerton College Cambridge. She has written numerous articles for curriculum journals, books for early years teachers and chapters in books on early years education and language in science. She is an assistant editor for *Primary Science Review* and her research interests include assessment of early years science. She continues to encourage many children's interest in science through her teaching and through leading BAYS Young Investigator clubs.

Philip Stephenson Philip has taught in both the primary and secondary sectors. He was both an Advisory Teacher for Science and a head teacher before joining Homerton College Cambridge as a Lecturer in Science Education. He is currently the Primary Science Co-ordinator at the College and is the Associate Director of the SCIcentre (National Centre for Initial Teacher Training in Science). He has written numerous primary science curriculum materials, a chapter on information technology and authentic learning, and material for the *Journal of Museum Ethnographers*. His research interests include work on science education for the able child and on Key Stage 2/3 cross-phase liaison.

Paul Warwick Paul is a Lecturer in Primary Science at Homerton College Cambridge. Prior to this he taught in primary schools in Cambridgeshire and was an Advisory Teacher for Science. At Homerton Paul is involved in teaching across undergraduate and postgraduate courses, while his research interests have focused on IT in science, children's understanding of the procedural aspects of science and the assessment of children's learning. He has written articles for curriculum journals and a study guide for data logging in primary science, and has contributed chapters to books on teaching and learning in the early years and on language in science.

Linda Webb Linda was Head of Science in a 13–18 school before becoming a Lecturer in Science Education at Homerton College Cambridge. While at Homerton she was involved in an NCET project evaluating the use of portable computers in schools and became increasingly interested in the use of IT to develop learning in schools. She taught both primary and secondary teachers and trainee teachers on a variety of initial teacher training and professional development courses. Among her publications are contributions to *Primary Science Knowledge and Understanding* (ed. Jenny Kennedy) and *Information Technology and Authentic Learning* (ed. Angela McFarlane).

Elaine Wilson Elaine was a secondary chemistry teacher for fifteen years. She was awarded the Salters' Medal for outstanding chemistry teaching in 1995 and was recently appointed to the post of Lecturer in Science Education at Homerton College Cambridge, where she has become increasingly interested in the science understanding of middle school pupils. She has published a range of curriculum materials to support the teaching of chemistry and has research interests in peer tutoring and learning styles in science teaching.

Series Editors' Preface

It is probably fair to say that science has been one of the successes of the implementation of the National Curriculum in the United Kingdom, especially for children in the 3–13 age group. Whereas most primary children's experience of science before the National Curriculum was dominated by the ubiquitous 'nature study', the breadth of children's experience is now considerably wider and their achievements equal to the best performances elsewhere in the world. This is in no small measure due to the quality of the researchers, writers and teachers of science in the United Kingdom.

All of the contributors to this publication have substantial experience of science in the primary and middle years as writers and researchers, teachers and teacher educators. Their contributions continue and sustain the high standards of debate and reflection achieved by science educators as they consider the most important issues related to the continuing development of the subject. Those involved in science education, more than in most curriculum subjects, have effectively linked the findings of theory to practice which have ultimately led to changes in teachers' thinking and the ways in which they work in the classroom.

This volume in the Primary Directions Series continues this important contribution to science and to education more generally by providing a thoughtful and constructively critical review of the past, present and future of the subject. It will be of relevance to a wide range of readers, to students in training, to teachers on in-service courses, to lecturers, researchers and to policy-makers – in fact, all those who wish to understand the history and principles upon which science has developed for children aged 3–13 and the ways in which this will continue in the future.

Colin Conner and Geoff Southworth

Acknowledgements

The editors wish to thank Richard Gott and Sandra Duggan for granting permission to reproduce (in Table 5.1, p. 60) much of the content of Table 2.2 from p. 31 of *Investigative Work in the Science Curriculum* (1995), and Alex Johnstone for granting permission to adapt a table from 'Why is science difficult to learn? Things are seldom what they seem', *Journal of Computer Assisted Learning* (1991) 7: 75–83, reproduced as Figure 4.1, p. 45.

1 Introduction

Rachel Sparks Linfield and Paul Warwick

The following caricatures of the place and nature of science in the curriculum at different points over recent decades are likely to stir the memories of many educators.

In the 1960s science did not feature greatly in the curriculum until pupils reached secondary school. In the primary school the 'nature table', sometimes dusty and given scant attention by pupils, was the main focus of attention for scientific endeavour. The criteria for objects' inclusion were that they were 'interesting' and had been found in the natural world, often by individual children on class walks. Science lessons, usually called 'nature study', tended to focus on this collection of objects, which were selected to be drawn, identified and described. Rarely were the items part of a planned whole school curriculum in which areas for study were matched for children's abilities or stages of development. Knowledge about the objects, frequently superficial, tended to dominate teaching as opposed to the development of the skills necessary to investigate. Thus, the picture of science in the typical 1960s primary class is a lesson focusing on knowledge about nature. Children often, for example, knew the colours of common birds' eggs but little about patterns or trends in birds' habitats or life cycles. They might know that a blackbird's egg was blue but be unable to recognise or give reasons for the features of the bird itself.

The Plowden Report emphasised the importance of child-centred education. It promoted the need for a curriculum characterised by 'activity and experience, rather than of knowledge to be acquired and facts to be stored' (Plowden, 1967). This led to the encouragement of group work, discussion and problem-solving. Science, where taught, began to be linked to a class topic. This greater diversity in content was encouraging and in some schools began to be supplemented by a concern to teach science skills. Yet topics such as 'the Vikings', 'Water' and 'the Sea' would all lead to experiments on floating and sinking regardless of the ability or age of a class. A progressive curriculum was rarely promoted, and therefore it is of little surprise that a survey of primary schools carried out by the Department of Education and Science (DES) in 1978 found that few of the schools they visited had 'effective programmes' for teaching science.

During the same period the science curriculum for the 11–13 age group could be characterised in several ways. Little or no account was taken of pupils' previous experiences. They were treated as 'blank slates', ready to receive the truth of science through a combination of didactic teaching, demonstration, practical work providing sometimes dubious evidence for the conclusions that were required to be drawn, and dictation of science facts. No premium was placed on the ability to articulate thinking about the concepts or procedures of science.

Two decades later, the implementations of the National Curriculum in England and Wales, the 5–14 National Guidelines in Scotland, Northern Ireland Curriculum and changes in the structure of examinations have had profound effects upon the way science is viewed and taught. In addition, educational research, the influence of professional bodies and the way teachers are trained have also had major ramifications for the development of science education, which has a higher profile than ever.

To what extent might such impressions be supported by the evidence? This book considers the factors which have been influential in developing science teaching and learning for the 3–13 age range over recent decades. We investigate the impact these factors are likely to have on the future direction of science teaching and learning in the twenty-first century. Whilst historians plot the course of major events in the development of science education, as science educators we hope to reflect on what has happened, to consider where present trends are likely to lead and whether this direction is, in fact, desirable. We have chosen to do this by considering particular features of, and priorities within, science education over recent decades.

Throughout the decades from the 1960s to the late 1980s the 'content versus process' debate informed discussion, enquiry, research and curriculum development in science. This debate reflected the complex interrelationships that exist in the consideration of any single aspect of science education; in discussing one aspect, others were naturally encompassed in the discussion. Within this book, the decision to assign chapters to specific aspects of 3–13 science has a similar effect to that experienced with the content versus process debate – no matter what the aspect under consideration, there is inevitable overlap in the material being discussed. Themes such as the value of practical, investigative science are considered by a number of authors. Bodies such as the Assessment of Performance Unit (APU) and researchers such as Driver and Harlen are cited in almost every chapter. It seemed important to us to give readers the option of focusing on a specific topic if they so wished, yet the overriding intention in structuring the book in this way was to allow ideas to be revisited from a number of perspectives. It is hoped that this will build a more complete picture of how 3–13 science has evolved and provide wider views on its possible futures. It is, after all, this close interrelationship of themes and changing perspectives that makes science education such an exciting area in which to be involved.

Ian Morrison and Linda Webb begin, in Chapter 2, by considering the development of primary science and the impact of the National Curriculum.

Despite the protestations of teachers and the practical problems surrounding its initial implementation, there can be little doubt that the National Curriculum did improve the status of science, making it one of three core subjects within the primary curriculum. Recently, initiatives to increase time spent on literacy and numeracy in school have highlighted concerns over the likely decrease in interest and time given to science. This is tempered, however, by the fact that trainee primary teachers are encouraged to spend an increasing amount of time and energy on improving their own levels of scientific knowledge and understanding.

Chapter 3 explores the growth and development of early years science within the context of a changing curriculum in which numeracy and literacy are increasingly specified in terms of content and time. Through considering the characteristics of good practice and the demands of recent legislation for early years education, Penny Coltman emphasises the need for practical, 'hands on' science and a genuinely early years curriculum for science.

Research into children's ideas in science has led to increased understanding of the way children view the world and the ways in which teachers can help children's understanding to develop further. Where once children's prior scientific knowledge and understanding was ignored, teachers do now try to differentiate and to take account of what children already think. In Chapter 4, Elaine Wilson indicates that there is still some way to go, though more thought is now given to what children *understand* as opposed to what they can simply *remember*.

This aspect of understanding is further highlighted in Chapter 5 which explores ideas about procedural understanding, the 'thinking behind the doing' (Gott and Duggan, 1995). The emphasis given within the National Curriculum to the development of process skills and the investigative approach can be viewed as having a long pedigree. Curriculum initiatives and research programmes focused on the scientific way of working have been instrumental in forming and re-forming attitudes and practice. Yet there remains much to do if *understanding* is to be given as much emphasis here as is now the case with respect to science concepts.

In Chapter 6, consideration is given to the assessment of learning in science. Again, the emphasis is on pupils' understanding and the role of formative assessment in moving learning forward. Initiatives to develop these links are explored and there is a critical examination of the possibilities for a future in which the use of IT is embedded in science activities. Assessment techniques and research into assessment are discussed. The constraints effected by the introduction of end of Key Stage testing are considered and the future is considered in the light of the diagnostic assessment that is already undertaken within the other core subjects of the National Curriculum.

Views on the 'right kind of teacher' for 3–13 science are tackled by John Hobden in Chapter 7. The changing nature of what it has meant to be the right kind of teacher, and whether such a concept is indeed helpful in developing science education, is explored in the light of research and curriculum development. Teacher training, both initial and in-service, is seen as a key to developing the role of the teacher of science and the possible role of the science 'specialist' in the primary school is given careful consideration.

There is a strongly implied link between this work and Angela McFarlane's chapter on the impact of IT (Chapter 8). Here, the relative paucity of effective links between science teaching and IT in recent decades is examined. Initiatives to develop these links are explored and the possibilities for a future in which the use of IT is embedded in science activities are examined critically. In an increasingly technological age, it may be that teacher expertise and willingness to innovate can be developed but that other constraints may prevent full advantage being taken of the technological possibilities.

In Chapter 9, Philip Stephenson examines the influence of interest groups, both professional and external to the education community, with respect to the developing science curriculum. The range of support available for those teaching science to younger children is made clear and the motivations of various interest groups are examined. Significant attention is given here to the Nuffield Foundation's publication *Beyond 2000* (Millar and Osborne, 1998) which considers the nature of science education for the future.

This consideration is central to Chapter 10, 'Science in Society or Society in Science?' Examining the relationship between science and society, Michael Reiss looks at the ways in which teachers view science education and the effects that these views might have on pupils' experience of the subject. In particular, he considers how science teaching might combat stereotypical views of minority groups within society, giving everyone a sense that science holds something for them.

What Are We Hoping For?

We recently asked a group of children from a range of backgrounds whether science was important. Their responses included:

> It's fun. We get wet sometimes. I don't like doing it on the blue table but I do like frogs and these things [magnets]. Do you want to see my bean? It's got woots. Kirsty's got leaves but I like my woots better. Come on!
>
> (Conor, aged 3)

> We do 'speriments in science. We find out 'thinks' like did you know plastic would be good for an umbrella 'cos it don't let water through? Science is really 'portant.
>
> (Ella, aged 5)

> It's useful for helping us know things. I like science, even when I have to write it all up. There's so much to find out. I mean, I know how to use a motor and what light bulbs do but I've still not seen the elections. I think they come out of the battery and move around the wires. Daddy says there's going to be local elections next week. I hope he'll save some to show me!
>
> (Ben, aged 8)

> Science helps us to know what to do. A lot of people are alive today who might have died if it were not for science. I like doing experiments. It also helps me to be concise. I mean, when I write up experiments I use sub-headings and things like that. In English I think I write better because I have to be accurate in science.
>
> (Emma, aged 13)

All of these responses demonstrate the genuine desire to discover more that is characteristic of many of those who have engaged, as teachers or learners, in the last thirty years of science education for the 3–13 age range. Clearly, we would hope that such positive attitudes are perpetuated. By considering the influences of the past and present outlined in these pages, we should be better able to promote the positive features of possible futures for science 3–13, bringing them closer to become part of a future reality.

References

Central Advisory Council for Education [England] (1967) *Children and their Primary Schools* (The Plowden Report), London: HMSO.

Gott, R. and Duggan, S. (1995) *Investigative Work in the Science Curriculum*, Buckingham: Open University Press.

Millar, R. and Osborne, J. (1998) *Beyond 2000, Science Education for the Future*, London: Nuffield Foundation.

2 The Development of Science 3–13 and the Impact of the National Curriculum

Ian Morrison and Linda Webb

To understand the growth of science in the curriculum for pupils in the age range 3–13 it is necessary to place the changes within the framework of changes to schools, to their teachers and to the curriculum taught. It is a transition from autonomy and freedom to control, prescription and accountability, 'from arguments about whether science ought to be taught at the primary level to how it should be taught' (Harlen, 1993).

Such changes have altered the balance of the primary school curriculum and seriously affected the teachers, and especially those trained before the changes were put in place. This introduction will initially focus on the overall changes, exemplified from science as appropriate, and will examine the key issues and influences on the development of science teaching up to the present.

A starting point must be chosen and it is convenient to go back to the 1960s, when schools and teachers had a large measure of control over the curriculum. The 1944 Education Act had said little about the curriculum. In general, governments had not wished to centralise control of education, a result of observing totalitarian governments overseas before and during World War II. With the 1960s came the beginnings of change. The Minister of Education, Sir David Eccles, wanted more government involvement, and a study group was set up on the curriculum. In 1964 this evolved into the Schools Council for the Curriculum and Examinations, which was not representative of the government alone but included local education authorities (LEAs) and the schools each as equal partners with central government. Other influences and changes were to follow with increasing frequency. The era began with considerable autonomy for schools and individual teachers. By the late 1990s there was prescription of what to teach in all subjects, including science, and prescription of how numeracy and literacy were to be taught, schools and teachers being accountable for the outcomes.

Early Influences

In the 1960s, teachers were encouraged to change their ideas and practices through curriculum development funded by the Nuffield Foundation and Schools Council projects, e.g. Nuffield Junior Science (1964–66) and Science

5–13 (1967–75). Both emphasised child-centred teaching approaches which acknowledged and valued children's ideas. This has been continued to the present day with the publication of Nuffield Primary Science (1995) based on the SPACE research (1990); for example, the *Teachers' Guide* for the topic 'Light' (Nuffield, 1995) was developed from the earlier research report on the same topic (Osborne et al., 1990). Second, the introduction of a comprehensive system of secondary schooling ended selection at the age of 11 years in many places, and this removed a major influence on the primary curriculum. Coincident with these was the publication of the Plowden Report (CACE, 1967). Diane Hofkins, writing on the thirtieth anniversary of its publication, says that

> The power of the report, which promoted 'child-centred' learning, where 'finding out' proved better for children than 'being told', is that after so long it remains an icon for both disciples and opponents. Reading Plowden today can induce a longing for that lost golden age in Britain, one that probably never existed, where there was such a thing as society, and everyone believed in pulling together to solve the ills of poverty, inequality and boring primary schools. In its sparkling optimism, and touching belief in social engineering, Plowden was very much of its time.
>
> (Hofkins, 1997, p. 2)

The report contained many ideas which are now being put in place, such as nursery provision, limiting class size for infants to thirty pupils, and raising the status of teachers with new in-service qualifications. There is consideration of 'How Primary Schools are Organised' (CACE, 1967, Ch. 20) with an advocacy of the inclusion of individual and group teaching as well as class learning. The report suggested that group work had often demonstrated the capacity of primary school children to plan and follow up mathematical and scientific enquiries. It was seen that pupils are more likely to propose a hypothesis to a group when they get the chance to discuss a problem and so understand that problem more clearly. This change of classroom practice helped open up the curriculum to more science, and the description above is reflected in the way science is often conducted in the primary classroom.

However, the child-centred approach, often ridiculed with the word 'progressive', is now condemned by 'sound-bite' politicians and journalists. Lady Plowden, interviewed thirty years after the publication of the report, without hesitation described 'good Plowden' and 'bad Plowden' schools. 'In the first, while the focus is on the individual child, the teachers keep absolute control and discipline is good. In the second, teachers "lose their heads"' (Passmore, 1997).

The Years of Disquiet

Within ten years of the Plowden Report, the mood had changed within education and in the country as a whole. There was a growing demand for accountability in education and for control of the curriculum. The Black Paper series, first published in 1969, the oil crisis and economic problems of the mid-1970s and the inquiry into the William Tyndale School (Auld, 1976) were the background to the Ruskin speech, made in 1976, by the then Prime Minister, James Callaghan. He called for a public debate on education, and one which was not just confined to the professionals in education. The tenor of the strongly critical comments in the press and elsewhere on education and educational standards can be found in para. 1.2 of *Education in Schools: A Consultative Document*, presented to Parliament as a Green Paper.

> Teachers lacked adequate professional skills, and did not know how to discipline children or to instil in them concern for hard work or good manners. Underlying all this was the feeling that the educational system was out of touch with the fundamental need for Britain to survive economically in a highly competitive world through the efficiency of its industry and commerce.
>
> (DES, 1977, p. 2)

The Green Paper argued that it was untrue that there had been a general decline in educational standards, but the need for schools to demonstrate their accountability to society required a coherent and soundly based means of assessment for the educational system. This would be provided by Her Majesty's Inspectorate (HMI) and by the Assessment of Performance Unit (APU) of the Department of Education and Science. The latter's terms of reference included the development of methods of assessing and monitoring the achievement of children at school. By 1977 the APU was developing tests suitable for national monitoring in science.

HMI reported on primary schools (DES, 1978) and on secondary schools (DES, 1979). Tomlinson (1993) believes the primary report shows that the 'three Rs' were well taught, that reading standards continued an upward trend since the war and 'progressive' methods had not run wild – 75 per cent of teachers used mainly didactic methods, only 5 per cent mainly 'exploratory' and 20 per cent a combination of both. The deficiencies were not in the 'basics', and pupil learning was best where the curriculum was broad. Science was found to be badly taught or not taught at all, with an emphasis on biological topics and the acquisition of facts. A beginning of the promotion of the specialist teacher or curriculum leader can be detected, as it could in an earlier report on the teaching of language (DES, 1975). In science teaching, in the late 1970s and early 1980s, there was a debate on the development of process skills as opposed to just knowing the content of science. This debate is picked up again in the chapter on procedural understanding.

Science 3–13

Which factors have been influential in developing science teaching and learning for the 3–13 age group in the last twenty years? How might these factors have an impact on the future direction of science teaching and learning for this age range into the twenty-first century? How can teachers cope with the changes?

Science 3–13 explores some of the historical antecedents of the current position of science in the lives of younger children. It covers the various influences, from both within and outside the teaching profession, that have shaped the current science curriculum. Current practice is examined and, on this basis, speculations are made about the future position and direction of this important subject.

The contributors each cover a particular aspect of science for the 3–13 age range but common themes emerge, such as the influence of government intentions, particularly through the development of the National Curriculum. The role of research groups and the impact of IT on the development of science in schools are also examined. The views of the teaching profession as to what should be taught and how science and science teaching should be viewed within society are shown to be important factors in the mix that contributes to change.

This book forms part of a series of key texts which focus on a range of topics related to primary education and schooling. Each book in the **Primary Directions Series** will aim to review the past, analyse current issues, suggest coping strategies for practitioners and speculate on the future.

Paul Warwick is a Lecturer in Primary Science at Homerton College, Cambridge. He previously taught in primary schools and was an Advisory Teacher for Science. His research interests are IT in science, children's understanding of the procedural aspects of science, and the assessment of children's learning.

Rachel Sparks Linfield is a reception teacher at St John's College School, Cambridge, and was previously a Lecturer in Primary Science at Homerton College, Cambridge. She has written widely in the area of early years education, and her research interests include assessment of early years science.

Primary Directions Series
Series Editors: Colin Conner, School of Education, University of Cambridge, and
Geoff Southworth, Department of Education Studies and Management, University of Reading

Assessment in Action in the Primary School
Edited by Colin Conner

Primary Education – At a Hinge of History?
Colin Richards

Science 3–13
The Past, the Present and Possible Futures
Edited by Paul Warwick and Rachel Sparks Linfield

Science 3–13

The Past, the Present and Possible Futures

Edited by Paul Warwick and Rachel Sparks Linfield

London and New York

First published 2000
by RoutledgeFalmer
11 New Fetter Lane, London EC4P 4EE

Simultaneously published in the USA and Canada
by RoutledgeFalmer
Garland Inc, 19 Union Square West, New York, NY 10003

RoutledgeFalmer is an imprint of the Taylor & Francis Group

Typeset in Times by Taylor & Francis Books Ltd
Printed and bound in Great Britain by
TJ International Ltd, Padstow, Cornwall

British Library Cataloguing in Publication Data
A catalogue record for this book is available from the British Library

Library of Congress Cataloging in Publication Data
Science 3–13 : the past, the present, and possible futures / [edited by] Paul Warwick and Rachel Sparks Linfield
p. cm. – (Primary directions series)
Includes bibliographical references and index.
1. Science–Study and teaching (Elementary)–Great Britain.
2. Science–Study and teaching (Secondary)–Great Britain.
I. Warwick, Paul, 1956–. II. Linfield, Rachel Sparks. III. Series.
LB1585.5.G7 S33 2000
500'.71'2041–dc21 99–049509

ISBN 0–415–22786–0 (hbk)
ISBN 0–415–22787–9 (pbk)

This book is dedicated to our colleague Linda Webb (1948–98) who sadly died whilst it was being written

The Initiation of Change

Curriculum Development

From the beginning of the 1980s, the government wanted to be more involved, to have knowledge of the curriculum and, indeed, to influence the curriculum. The Green Paper *Education in Schools* (DES, 1977) emphasised the partnership between central government, the local education authorities and the teachers, and included the following in the section on 'Proposals and Recommendations'.

> The design and management of the school curriculum play a central part in determining what is achieved by our schools ... The Secretaries of State propose a review of curricular arrangements, to be carried out by local education authorities in their own areas in consultation with their teachers. They will in the light of the review seek to establish a broad agreement with their partners in the education service on a framework for the curriculum, and on whether part of the curriculum should be protected because there are aims common to all schools and pupils at certain stages. These aims must include the achievement of basic literacy and numeracy at the primary stage.
>
> (DES, 1977, p. 40)

This led to a series of publications on the curriculum in the first half of the 1980s from the DES, from Schools Council and from HMI. Throughout, there is still the notion of partnership between schools and LEAs with encouragement to analyse the curriculum and define aims. The DES document *The School Curriculum* (1981) clearly says that neither the government nor the LEAs should specify what the schools should teach, implying a non-interventionist approach. Then in 1984 the Secretary of State announced that he intended to seek broad agreement about the objectives of the 5–16 curriculum. HMI (DES, 1985a), in *The Curriculum from 5 to 16*, said that it was for each school to decide how the curriculum was to be organised. They reported that six general aims had received widespread support and been reflected in the aims drawn up by many LEAs and individual schools (DES, 1981). These aims were:

- to help pupils to develop lively, enquiring minds, the ability to question and argue rationally and to apply themselves to tasks, and physical skills;
- to help pupils to acquire knowledge and skills relevant to adult life and employment in a fast changing world;
- to help pupils to use language and number effectively;
- to instil respect for religious and moral values, and tolerance of other races, religions and ways of life;
- to help pupils to understand the world in which they live, and the interdependence of individuals, groups and nations;

- to help pupils to appreciate human achievements and aspirations.

(DES, 1985a, p. 3)

Some of these aims can be related to the requirements of National Curriculum Science. For example, Key Stage 1 pupils should be given opportunities to ask questions such as 'How?', 'Why?' and 'What will happen if ...?'

Clearly, such questions can be easily linked to the first two aims quoted above. Good infant teaching will encourage the lively enquiring minds by raising all sorts of questions with the focused exploration placed in contexts which are real to the pupils.

Within *The Curriculum from 5 to 16*, the section on the teaching programme uses the word 'may', as in 'the teaching programme may be planned in a variety of ways'. There is reference to the use of role play and constructional toys for younger children and to themes or topics closely linked to older primary pupils' experiences and interests. The value of such work is expounded at length, but concludes with a short but nearly prophetic caution, prophetic in the light of what was to follow in the next years.

> It can be difficult to ensure that there is sufficient progression and continuity, particularly for older children, in the work in each area covered by a topic, and it may be easier to plan progression if some of the work is organised in separate subjects.
>
> (DES, 1985a, p. 9)

HMI proposed that the curriculum of all schools should involve nine areas of learning and experience, not discrete subjects, complemented by four elements of learning, the knowledge, concepts, skills and attitudes (DES, 1985a). Schools were to establish the extent to which particular topics, aspects and subjects are already contributing to these areas and to the development of the four elements of learning. These were seen to be planning and analytical tools. An example of the approach is the idea that scientific learning introduces practical experiments as a means of investigating observed phenomena, while offering valuable opportunities to develop more general skills such as approaching tasks in a logical manner, communicating information and ideas, and observing and recording. A single activity was seen to be able to contribute to several areas of learning. This topic approach influenced the development of the 3–13 curriculum and was later decried by the Black Papers and the Assessment of Performance Unit (referred to in detail in the next section) because it did not lead to effective, coherent science teaching in primary schools. The latter's survey of schools (APU, 1988) had asked schools to select the five goals they considered most important. It was found that the most highly rated goals were often the general goals of primary education, while the least rated goals were those specific to science activity.

By contrast to this approach, HMI (DES, 1985a) said that 'in secondary schools subjects are well established as a convenient and familiar way of organising learning, and in their selection of subjects they try to ensure for their pupils a broad, balanced and useful education'. So the ease of planning in terms of subjects rather than the areas of experience was noted. Throughout, there is the message that responsibility for the details of the curriculum belonged to the school in partnership with the LEA.

Science, the Primary Curriculum and its Teachers

It is possible to identify three main influences on the teaching of science during the 1980s. These are (i) *Science 5–16: A Statement of Policy* (DES, 1985b); (ii) the national system of assessment developed by the APU; and (iii) the promotion of 35-day Inset courses in science for primary teachers.

(i) Science 5–16: A Statement of Policy

This policy statement underpins the other two influences and builds on reports published earlier, including those by the Association for Science Education and the Royal Society. The emphasis is on 'Science for all', and the statement promotes the values of the processes of science. A key issue in the science chapter in *Aspects of Secondary Education in England* (DES, 1979) had been the balance between content and process.

> In many schools, the acquisition of knowledge was the main feature of the science courses; only rarely was the emphasis on teaching the process of science rather than the subject matter. There were few situations where an appropriate balance between content-based teaching and the 'science as a process' approach had been achieved.
>
> (Para. 12.4)

The authors of the policy statement believed an essential characteristic of science education was the introduction of the methods of science. There is a list of nine features, very similar to those in Science Attainment Target 1: for example, 'seek and identify patterns and relate these to patterns perceived earlier'. This is explored further in the chapter on procedural understanding.

The document also reviewed the provision of science in primary schools using evidence from the APU reports and from HMI. Progress is acknowledged, but there is regret that the initial positive impact of the Nuffield Junior Science Project and the Schools Council's Science 5–13 Project, and the considerable influence on the participating teachers, had not been sustained. HMI surveys of primary education in the late 1970s had shown that only a minority of schools possessed effective programmes for teaching science. However, even by 1983, the substantial efforts by LEAs, the DES, the Association for Science Education and the schools themselves were evident in the APU survey (DES, 1983) of primary science, which showed that 90 per cent of schools sampled said they included 'science activities in the curriculum'. An increasing number of curriculum publications during the 1980s emphasised a structure based upon appropriate content rather than simply on activities linked to the development of broad investigational skills. To consolidate this progress, the Secretary of State asked LEAs to develop and publish policies for science education and plans for implementing them. Experience showed that for plans to succeed, schools needed to have at least one teacher with 'the capacity, knowledge and insight to make science education a reality' and that 'the objective should be that all class teachers, without exception, should include at least some science in their teaching' (DES, 1985b). In retrospect, it can be seen that the policy statement was laying the foundations for science as a core subject in the National Curriculum, not just included as an optional activity.

The greatest perceived obstacle to continued improvement was the poor knowledge of elementary science held by many existing teachers, no matter how creative and innovative they were. The policy statement proposed action in initial teacher training (ITT) and in in-service teacher training (Inset). In 1984 the DES issued the first circular (DES, 1984a) for the approval of ITT courses, setting out criteria against which all courses would be assessed. No reference was made to science in these 1984 criteria. However, in the 1985 policy statement there was an expectation that all ITT institutions would 'review the adequacy of their specialist staffing in science, so that all new primary teachers may be provided with a firm foundation in the subject, and some may be equipped to act as curriculum leaders'. By 1989 the criteria for curriculum studies in primary courses required that at least 100 hours should be devoted to the teaching of mathematics, to the teaching of English, and the teaching of science (DES, 1989b). A version of this criterion has remained in place ever since.

Inset was also seen as a vehicle for continued improvement, and primary science courses were included from the academic year 1984–5. The courses consolidated the teachers' scientific knowledge, giving them the confidence to teach science and enabling them to practise scientific skills and methods. A feature was the inclusion of physics topics, such as forces and measurement, electricity and energy. Many teachers found some of the subject matter difficult, especially as they felt that they should just be taught how to teach primary science and not have their own knowledge raised as they practised their scientific skills. Teaching the teachers was stimulating, as the cohorts went from a feeling of being pioneers in the early years to recognising the demands being made of them when the National Curriculum was first published.

(ii) The National System of Assessment Developed by the APU

The Assessment of Performance Unit of the DES was formally established in 1975 and referred to in the Green Paper. The model of the curriculum originally proposed was based on six lines of development which crossed subject boundaries, and was not dissimilar to the nine areas of learning and experience found in *The Curriculum from 5 to 16*. The proposals and recommendations in the Green Paper (DES, 1977) said that the APU would concentrate on the development of tests suitable for national monitoring in English language, mathematics and science. In other words, the focus was already on what we now call the core subjects, a very different curriculum model.

The framework for assessment in science reflected the variety of activities which contribute to performance in science, and the test items were described according to three characteristics: scientific process, science concept and context. The process list of six categories, the assessment framework, defined key aspects of science teaching, constituted good practice in teaching science and can be found today in Attainment Target 1: for example, using apparatus and measuring instruments, and observation tasks. The framework is as follows:

1 use of graphical and symbolic representation;
2 use of apparatus and measuring instruments;
3 observation;
4 interpretation and application;
5 planning of investigations;
6 performance of investigation.

(APU, 1988, p. 2)

Pupils were tested at age 11, the top primary year, and at age 15, near the end of compulsory schooling. In addition, 13-year-old pupils were included, because most maintained schools taught 'science' up to that age and then split the sciences afterwards.

The results were published in reports, and these and the test equipment were used in the 35-day courses for primary and secondary teachers. So the basis of the testing became a vehicle for Inset courses and could be seen to have acted as a starting point for the National Curriculum. The outcomes from the APU surveys 1980–4 were summarised (APU, 1988) and communicated in a series of briefer reports for teachers (DES, 1983, 1984b, 1986). They all contained sections on the implications for teaching.

For example, pupils at age 11 were willing to be involved in investigations, but there was little evidence that they considered their work in a critical, reflective way. While differences between the sexes were not marked, girls were slightly ahead in using graphs, tables and charts and in making and observing similarities and differences. Boys were ahead in using measuring instruments, in applying physical concepts to problems and in recording quantitative results in investigations. Pupils did better in the general skills important to all parts of the curriculum and less well in the skills more specifically related to science activities. Too often, potentially rich science experiences were narrowed to observing and recording.

> There is now a need to consider how to help children acquire those more specific science skills such as defining patterns in observations, giving explanations, predicting, hypothesising, controlling variables and planning investigations ... Planning must also take account of the teaching approaches which will be used to develop science process skills, since they can only be acquired if the children are given opportunities to use them.
>
> (APU, 1988, p. 9)

This quote is from the *Science at Age 11* report (DES, 1983) and was repeated in the *Science at Age 13* report (DES, 1984b), with the statement that many of the process skills in which performance was low appear to be the basic needs of science courses of all kinds. In a special report on planning scientific investigations at age 11 (DES, 1986), the theme continues, with 'It is unsurprising that children are unable to produce an adequate plan as it seems they often have little opportunity of finding out how to do it.' The APU strongly encouraged teachers to plan activities which enhanced pupils' science abilities in terms of precise processes and concepts. This can again be seen as a forerunner of the planning needed after the introduction of the National Curriculum in Science.

In addition to these three major influences, at the end of the decade HMI published a review of *The Teaching and Learning of Science* (DES, 1989b). It drew upon inspections of nearly 400 primary schools, and the findings were within the framework of the publications cited above. It was written at the time of the first reactions to the National Curriculum. It highlighted a number of weaknesses.

> Many primary schools approach the teaching of science through topic work which involves aspects of other subjects, such as history or geography. In the best circumstances this approach provides scope for developing children's knowledge, skills and understanding in science. However, where such topic work is not well planned, or where too many aspects of different subjects are attempted, the work often lacks coherence and as a result the children receive a superficial experience of science.
>
> Few schools effectively assess children's progress in science. This often results in weaknesses in matching the work to the ability of the children and in planning progression in science within and across year groups.
>
> (DES, 1989b, p. 5)

Progress was recognised, but the conclusion was that the majority of primary schools still needed to improve the balance and breadth of their work and give more attention to assessing and recording children's progress. The early reactions to the National Curriculum in Science indicated that some primary schools found the Attainment Targets and Programmes of Study helpful in overcoming these difficulties.

The National Curriculum

The scene changed when Kenneth Baker was Secretary of State. He saw muddle where others saw variety, a result of the state handing over its responsibilities to the other two partners, the LEAs and the schools and individual teachers. Central government was going to prescribe the curriculum, ending devolved partnership arrangements. Even before legislation was in place, working groups were set up on mathematics and science, advising the government on the Attainment Targets suitable for children at four key ages, and the Programmes of Study which would allow them to be achieved. The DES published *The National Curriculum 5–16: A Consultation Document* in July 1987. The curriculum was described in terms of academic subjects, so very different from the areas of experience described in the HMI publication *The Curriculum from 5 to 16* (DES, 1985a). Primary school teachers, more used to topic work and themes, had found the 1985 model to their liking and strongly disapproved of the subject-based curriculum, its prescription and assessment.

Science and the National Curriculum

The major benefit of the National Curriculum has been the raised profile of science as a core curriculum subject, so very different from the earlier inclusion of science activities in the curriculum. Black (1990) believed that a major achievement of APU Science was to establish the process approach within school science and to make a critical contribution to the National Curriculum Council's work in this area. The debate over whether the emphasis should be on 'process' or 'content' continued during the early years of the Science National Curriculum. Swatton (1990) believed that dogmatic adherence to one or other side would lead only to further confusion among pupils and teachers over what science is all about, and in particular the role and status of practical work in science lessons. This argument now seems to be given much less prominence, and Swatton believed teachers generally adopted an 'holistic view', with a much more organic relationship between process and content in children's learning. If this had not happened, the role of practical work in school science would have been unclear and the pupils' achievements unrecognised.

The APU had emphasised the importance of practical testing in science monitoring, and Black (1990) believed that 'monitoring without the extensive use of practical assessment could do serious harm to school science'. However, after initial attempts to carry out this strategy, immediately after the inception of the National Curriculum, testing became short, cheaper and more like traditional tests. Later, Black (1995), in reviewing the implementation of the National Curriculum for science said that:

> For many who had never planned topic work with clear aims in mind leading to work that could serve to assess pupils' learning, they [the initial tests] were both demanding and disorienting. Many primary teachers complained about the unfamiliarity and the onerous workload, and received much media attention.
>
> (Black, 1995, p. 176)

The development of the National Curriculum in Science can be traced by looking at back copies of journals such as *Primary Science Review*. The editorial for Autumn 1987 (Roberts, 1987) is entitled 'Our response to new initiatives'. Writing when the Education Reform Bill of 1988 was imminent, the editor asked what were the current staffroom discussions. They probably included the centrally controlled National Curriculum, and testing at age 7. One year later, the editorial was 'Getting to know the National Curriculum'. Here, Harlen (1988), as a member of the working group on the science curriculum, believed a framework had been provided which supported teachers, reflected good practice and met the requirements of the Secretary of State. This editorial also provided teachers with informative detail on the Programmes of Study (what is taught), the Attainment Targets (the learning objectives), assessment and implementation.

In the summer of 1989 the Association for Science Education published a special supplement on the National Curriculum. The very practical articles included 'Assessing the National Curriculum', 'Support for teaching' and 'Preparing and planning in school'. There was information, such as the fact that three of the seventeen Attainment Targets were not applicable for pupils aged 5–11 years. The article, 'An ASE primary science perspective on the National Curriculum' (Ruddle, 1989), has an interesting insight into the development of science teaching in the primary school in the late 1980s. The Association's view had been that science should be taught by all class teachers in a primary school, but it acknowledged that this had still not been accepted by everyone. Many teachers felt ill prepared to teach science, and the mass of detail in the science document daunted them.

The National Curriculum challenged teachers to provide continuity and progression, two of the principles for science education already proposed within *Science 5–16* (DES, 1985b). School policies had to be reformulated by head teachers and staff, with support from LEA advisers and from the ASE itself. Primary and secondary schools were required to have curriculum discussions, so that the secondary school paid attention to the foundations already laid in the primary school. This would enable them to give pupils progressively deeper understanding.

For the next two years, the *Primary Science Review* regularly visited issues of importance to the class teacher. However, even by 1990 there was a growing feeling that the broad and balanced curriculum as described by the National Curriculum was unmanageable. Difficulties occurred over the end of Key Stage assessment arrangements, with evidence from teacher assessment and the tests providing different results for many pupils. This led to the radical revision of the whole curriculum as published in the Dearing Report (1994). The curriculum was slimmed down and the assessment arrangements simplified. There is teacher assessment at the end of each Key Stage, and there are tests in the three core subjects, taken at the end of Key Stage 2. The results are now published for each school.

In Science, the seventeen Attainment Targets were reduced to four, setting out the expected standards of pupils' performance. These standards were then set out in eight Level descriptions of increasing difficulty, an innovation. The Programmes of Study again set out what pupils should be taught. However, each Key Stage (KS) Programme of Study had introductory requirements. These were: Systematic Enquiry; Science in Everyday Life; the Nature of Scientific Ideas; and Health and Safety. They applied as overarching requirements across the other four sections of the Programme of Study, and tended to be known as Sc0, with Sc1 being Experimental and Investigative Science, Sc2 Life Processes and Living Things, Sc3 Materials and their Properties, and Sc4 Physical Processes. (More recent revision has led to the renaming of Sc1 as Scientific Enquiry and the subsuming of elements of Sc0, for example Health and Safety, within the rest of the Programme of Study.)

The Dearing review not only slimmed down the curriculum, but explicitly changed the place of science in the National Curriculum. The three core subjects are no longer of equal weight, if they ever were. When it was initially introduced, it was expected that the National Curriculum would take no more than 70 per cent of the timetable, but schools, and especially primary schools, immediately complained of curriculum overload and the impossibility of fitting in all that was required. In the new arrangements, the number of hours per year to be spent on science at KS1 was 54, with 72 at KS2. This compared with 216 and 180 for English and 126 hours at both Key Stages for Mathematics. It was expected that some 20 per cent of time would be freed from curricular prescription.

The time allocated to science at KS1 was less than what is needed, and represented only an hour and a half per week. With National Curriculum testing focusing on the core subjects, it was felt that the 'discretionary' time would be used for the core subjects, and for information technology, which had been upgraded to a new subject.

These changes were implemented in August 1995, with a moratorium on further change for five years. By 1998, teachers and subject associations were being invited to comment on the need for, and the nature of, modifications to the post-Dearing curriculum. There were hints that science would not be changed radically. Indeed, Summers and Mant (1998) recognised a reluctance in primary school teachers to contemplate further change. They advocated that minor changes should occur where there were good arguments and research evidence to support the improvements. For example, they believed that the omission of energy as a topic at Key Stage 2 had been misguided. Energy is seen to be a concept fundamental to many others, and they proposed that it could replace some of the existing work on balanced and unbalanced forces and their link with motion. They also believed that research showed that some additional concepts could be introduced into the teaching of electricity, as they are readily accessible to children at KS2, and that these could be linked with energy. In fact, the changes made to content were relatively minor. Pleasingly, more prominence was given to individual needs of all children through an emphasis on inclusion.

The success of primary science since 1989 was documented by Ponchaud (1997) in his role as the Ofsted subject adviser for science and drawn from the data provided by 13,000 lessons observed in 1995–6. He reported that pupils made steady gains up to the end of KS1, progress was slower in Years 3 and 4, and that pupils' knowledge and understanding of important ideas in science greatly improved towards the end of KS2. The good overall progress reported from inspections was also reflected in national test results and teacher assessment, which showed that in 1996 more pupils reached or exceeded national expectations in science than in the other two core subjects. Further confirmation of the progress is found in the Third International Maths and Science Survey, which showed that the performance of English 9-year-olds compared favourably with those in most participating countries. In no Western European country were the mean scores of Year 4 or Year 5 pupils significantly better than those of children in England.

This Ofsted report also showed that pupils' attainment at the end of KS2 is highest in AT2 (Life Processes and Living Things) and substantially weaker in AT4 (Physical Processes), with particular difficulties occurring in relation to their understanding of forces and light. Some pupils achieved very high standards in AT1 (Experimental and Investigative Science), but attainment was more variable in this area. For example, it is recommended that pupils need to develop a better understanding of why investigations need to be 'fair' and what constitutes a conclusion from practical activity. More positively, most children were able to use simple apparatus correctly, make and record observations and identify patterns and trends, possibly benefiting from the work of the APU.

A Possible Future for Science

After their election victory in 1997, the Labour government wanted teachers to focus their efforts on the four subjects, English, mathematics, science and information technology. Daily literacy and numeracy teaching is to enable a much greater proportion of 11-year-olds to reach the higher required standards by 2002. The government has perceived that the National Curriculum has, to an extent, been getting in the way of teachers giving literacy and numeracy proper attention. Thus, from 1998, teachers were not required to deliver the detailed national syllabus in the foundation subjects.

This shift in emphasis has also occurred in primary Initial Teacher Training (ITT). Successive phases of inspections have focused on the training of students in literacy and numeracy, then on English or mathematics plus a student's specialist subject, and then on science as well as information and communications technology (ICT). There is a trend to a situation where English and mathematics are seen as the core subjects and ICT and science as subsidiary core subjects, but still very strictly controlled by the statutory orders.

For nursery education, goals have been written providing desirable outcomes for children by the time they enter compulsory education (DfEE, 1998). These outcomes emphasise early literacy and numeracy and the development of personal and social skills, and they contribute to children's knowledge, understanding and skills in other areas. Presented as six areas of learning, each desirable outcome is related to National Curriculum Level 1 and 2 descriptions in different subjects, including science. For example, within Personal and Social Development, it is desirable that pupils work as a group, are eager to explore new learning, and show the ability to initiate ideas and to solve practical problems. This sentence reverberates with the Plowden Report's reference to group work, which was discussed earlier in this chapter. This desirable outcome leads into the AT1 Level Description, 'They respond to suggestions of how to find things out and, with help, make their own suggestions' (DfE 1995, AT1, Level 2). A similar link is made between 'Knowledge and Understanding of the World' and another end-of-Key-Stage description for Science AT1. These explicit references in the document may help overcome the attitude that no real science can happen before the age of 5. Rosemary Feasey (1997) believes that early years science focuses on the 5–7 age group, and that there is little material or debate on science in nursery education.

These changes will all strengthen the place of science in the primary school. Indeed, one of the successes of the National Curriculum has been the continued progress made in the teaching of science. The Ofsted overview (Ponchaud, 1997) referred to pupils entering secondary school with a grounding in scientific ideas, language and skills. These successes were built on the work of the 1980s which enabled many teachers to see the introduction of the science curriculum as a natural continuation of what they had already been doing.

So where should we be going? Though there are still teachers who lack confidence in their own scientific capabilities, this is improving. It will seem dull to say that the current efforts should continue, but we must ensure the output of newly qualified primary school teachers:

- have an accurate understanding of scientific concepts and skills;
- assess young children's scientific development effectively;
- provide scientific learning experiences that stimulate and excite children and demonstrate the relevance of their learning to everyday life and welfare;
- promote positive and non-stereotypical perceptions of science and scientists.

(Jarvis et al., 1997, p. 28)

This quotation is taken from the aims of SCIcentre, the National Centre for Initial Teacher Training in Primary School Science. Their report focuses on the use of directed time for training such students, on the design of courses for specialist teachers and on the methods for training teacher-mentors. This training is now much more prescriptive than it would have been even ten years ago. Student teachers' knowledge is being audited and then enhanced. The effectiveness of taught sessions is being increased by requiring preparation and follow-up work from students. Coherent experiences within a developmental structure link school and the college-based work. Students are required to learn about children's conceptual development and to use this knowledge to assess and structure the pupils' learning.

If this growth in professionalism is required of student teachers, then it is also necessary for more experienced teachers. This structure for ITT is very similar to the emphasis of the science Inset courses of the 1980s, but it is now underpinned by the National Curriculum. Teachers are currently being trained for the literacy and numeracy hours, and there is relatively little science Inset.

The model being developed by the SCIcentre for student teachers may become the future model for in-service training. Targets have not been set for standards in science as they have been for the basics, although inspections have highlighted areas for further development. These may only be achieved if teachers are given more support.

Whatever changes occur in the science curriculum, the biggest change may be in the greater scientific expertise of the specialist teacher and of the general class teacher.

References

APU (1988) *Science at Age 11. A Review of APU Survey Findings 1980–84*, London: HMSO.

Auld, R. (1976) *The William Tyndale Junior and Infants Schools. A Report of the Public Inquiry*, London: ILEA.

Black, P. (1990) 'APU Science – the past and the future', *School Science Review*, 72: 258.

Black, P. (1995) '1987 to 1995 – The struggle to formulate a National Curriculum for Science in England and Wales', *Studies in Science Education*, 26: 176.

Central Advisory Council for Education (CACE) (1967) *Children and their Primary Schools*, London: HMSO.

Dearing, R. (1994) *The National Curriculum and its Assessment: Final Report*, London: Schools Curriculum and Assessment Authority.

DES (1975) *A Language for Life* (The Bullock Report), London: HMSO.

DES (1977) *Education in Schools: A Consultative Document*, Green Paper, Cmnd 6869, London: HMSO, pp. 2, 40.

DES (1978) *Science Progress Report. Assessment of Performance Unit*, London: HMSO.

DES (1979) *Aspects of Secondary Education in England: A Survey by HMI*, London: HMSO, p. 188.

DES (1981) *The School Curriculum*, London: DES.

DES (1983) *Science Report for Teachers. 1: Science at Age 11*, London: DES, pp. 30, 31.

DES (1984a) *Circular 3/84 Initial Teacher Training: Approval of Courses*, London: DES.

DES (1984b) *Science Report for Teachers. 3: Science at Age 13*, London: DES.

DES (1985a) *The Curriculum from 5 to 16. Curriculum Matters 2*, London: HMSO, p. 3.

DES (1985b) *Science 5–16: A Statement of Policy*, London: HMSO.

DES (1986) *Science Report for Teachers. 8: Planning Scientific Investigations at Age 11*, London: DES.

DES (1987) *The National Curriculum 5–16: A Consultation Document*, London: DES.

DES (1989a) *Circular 24/89 Initial Teacher Training: Approval of Courses*, London: DES.

DES (1989b) *Aspects of Primary Education: The Teaching and Learning of Science*, London: HMSO, p. 5.

DfE (1995) *Science in the National Curriculum*, London: HMSO, p. 2.

DfEE (1996) *Nursery Education: Desirable Outcomes for Children's Learning on Entering Compulsory Education*, Sudbury, Suffolk: DfEE Publications.

Donnelly, J., Buchan, A., Jenkins, E., Laws, P. and Welford, G. (1996) *Investigations by Order: Policy, Curriculum and Science Teachers' Work under the Education Reform Act*, Leeds: Studies in Education.

Feasey, R. (1997) 'Early years science – the Cinderella of science education?', *Primary Science Review*, 48: 4.

Harlen, W. (1988) 'Getting to know the National Curriculum', *Primary Science Review*, 8: 2–3.

Harlen, W. (1993) *Teaching and Learning Primary Science* (second edition), London: Paul Chapman Publishing.

Hofkins, D. (1997) 'A golden age when Utopia was a possibility', Primary Update, *The Times Educational Supplement*, January 24, p. 2.

Jarvis, T., McKeon, F., Stephenson, P. (1997) *Course Design in Primary Science Initial Teacher Training*, Leicester: SCIcentre, p. 28.

Nuffield Junior Science Project (1967) *Teachers' Guides*, London: Collins.

Nuffield Primary Science (1995) *Teachers' Guide for Ages 7–12: Light*, London: Collins Educational.

O'Hear, P. and White, J. (1993) *Assessing the National Curriculum*, London: Paul Chapman Publishing.

Osborne, J., Black, P., Smith, M. and Meadows J. (1990) *Primary SPACE Project Research Report: Light*, Liverpool: Liverpool University Press.

Passmore, B. (1997) 'Champion of active learning', Primary Update, *The Times Educational Supplement*, January 24.

Ponchaud, R. (1997) 'Your Ofsted inspection is over – what next?', *Primary Science Review*, 49: 9–11.

Roberts, K. (1987) 'Our response to new initiatives', *Primary Science Review*, 5: 2–3.

Ruddle, H. (1989) 'An ASE primary science perspective on the National Curriculum', *Primary Science Review*, Special Supplement, 12.

Science 5/13 *Units for Teachers* (1967–75) (26 titles) London: Macdonald Educational.

SPACE (Science Processes and Concepts Exploration) (1990–2) *Research Reports*, Liverpool: Liverpool University Press.

Summers, M. and Mant, J. (1998) 'Revision of the National Curriculum: A view from the PSTS project', *Primary Science Review*, 52: 12–14.

Swatton, P. (1990) 'Process and content in the National Science Curriculum', *School Science Review*, 72: 259.

Tomlinson, J. (1993) *The Control of Education*, London: Cassell.

3 The Growth and Development of Science in the Early Years

Penny Coltman

In a Year 1 class, on a typical afternoon near the end of the twentieth century, the children are busy exploring aspects of magnetism. A table is set out with a selection of magnetic toys: building blocks, magnetic letters on a baking tray, and a theatre whose characters are moved from beneath using magnetic wands. In another area of the room a group of children are busily making their own magnetic fishing game. Some children are taking magnets 'for a walk' around the room, discovering which materials are attracted to them, and a final group are working with their teacher, deciding how best to test which is the strongest magnet by picking up some brightly coloured discs. In the room there is a gentle buzz of activity. Children are talking together about their discoveries, sharing ideas and making decisions. There is autonomy, independence, purpose and focus. The teacher has a clear objective to the lesson, and is engaged in eliciting responses from children using productive questioning based on a sound understanding of the science of magnetism.

By contrast, in a fading sepia photograph, rows of tiny figures gaze unsmilingly from the cramped, uncomfortable tiers of a nineteenth-century classroom, in which learning was primarily by copying, rote and recitation. No moving about for these 5-year-olds, except perhaps a little drill or marching around the school yard. Opportunities to acquire scientific concepts were limited to the realms of nature study. Not the most stimulating or enjoyable forms of education for the very young, but, perhaps surprisingly, the tide was about to turn, as the foundations of learning through experience were already in place.

Learning through Experience: Early Exponents

Rousseau (1712–78) was one of the earliest idealists whose thoughts contributed to modern methodology. He wrote deploring the sterile and miserable educational process to which children of his day were so unhappily subjected.

> Love childhood, indulge its sports, its pleasures, its delightful instincts ... Why rob these innocents of the joys which pass so quickly, of the precious

> gift which they cannot abuse? Why fill with bitterness the fleeting days of early childhood, days which will no more return for them than for you?
>
> (Rousseau, 1911)

Rousseau believed in a childhood of liberty, in which children would be trained to become self-reliant. Direct relevance can be seen to the development of the philosophy which underlies much of the present-day science curriculum. Rousseau aspired to learning entirely through activities, sensory stimulation and first-hand experience. 'Teach by doing whenever you can, and only fall back on words when doing is out of the question' (Rousseau, 1911).

At the turn of the nineteenth century, Pestalozzi (1746–1827) echoed this need for the provision of experiences through which learning might be achieved, and developed a method of teaching based on principles of number, form and language, all to be encountered through the observation and handling of objects (Rusk, 1955). The 'object lesson' became a familiar aspect of Victorian schoolrooms, adopted initially in Britain by pioneers such as Robert Owen in his New Lanark Institution for the Formation of Character, opened in 1816 for the purposes of educating the children of cotton-mill workers.

Sadly, the promising ideals of the object lesson became dulled and diminished as their use in the curriculum became more widespread. In their least productive form they became little more than a memorising of facts, names and attributes, devoid of real meaning or purpose. Handbooks of lessons were published which did not require teachers to have any understanding themselves of the scientific properties or principles exemplified. First-hand experience, in many instances, became no more than the viewing of an artist's impression, and the development of language was reduced to a list of words to be learned by rote (Browne, 1991).

One of the most important forms of experience through which children in the early years gain and develop understanding is now generally recognised to be play. Through play, children explore and experiment, learning is extended and consolidated, experiences re-enacted. It is an essential part of the process of acquisition, accumulation and assimilation of knowledge. It is in the writings of Froebel that some of the earliest acknowledgements of the value of play can be found.

Ten years after Robert Owen opened his Scottish school, Froebel published his influential work *The Education of Man*. In this, he describes play as 'The purest, most spiritual activity of man at this stage', but qualifies this eulogy with the caveat that play should be guided lest it develop into 'aimless play instead of preparing for those tasks of life for which it is destined' (cited in Rusk, 1955). Froebel's play was an emulation of adult activities, carried out using symbolic 'gifts' which were the resources for key 'occupations'. These occupations included handicrafts for older children to develop dexterity and valuable technical skills. Angela Anning (1991) writes:

> The legacy of Froebel's belief that practical craft work – making up geometric designs with coloured wooden squares and triangles, paper-

> folding and weaving, stick-laying and construction, sewing and embroidery – leading to sensory and language training should be central features of the school curriculum is still in evidence in infant classrooms.
>
> (p. 10)

Rudolph Steiner (1861–1925) was an even greater exponent of the value of play. Indeed, a child who was reluctant to play would be a cause for great concern in a Steiner school. The provision for an inviting environment to encourage play is fundamental to Steiner's philosophy: 'Play is a more complete undertaking than work, for play uses all our faculties, work only some. In play we ourselves lead. In work we are being pushed' (Gardner, 1993). Like Froebel, Steiner advocated learning through imitation of adult activities. To the present day, teachers in Steiner classrooms complete their own handicrafts whilst children work on their own, as fine co-ordination is developed within aesthetic contexts. Dexterity and intellect were believed by Steiner to be inextricably linked:

> one has observed that someone who moves his fingers clumsily also has a 'clumsy' intellect with not very mobile and flexible ideas and thoughts, while he who knows how to move his fingers properly also has mobile and flexible thoughts and ideas which can enter into the essential nature of things.
>
> (Steiner, cited in Pusch, 1993, p. 125)

Colour, form and texture are explored as children carry out their crafts using natural materials, and indeed, the science of the Steiner Kindergarten is traditionally that of the natural world. Children are explicitly encouraged to develop a sense of wonder and appreciation in relation to their environment, with the resultant 'alternative' perception of the Steiner methodology.

The advocacy of learning through sensory experience found an influential exponent in Maria Montessori (1870–1952), who worked with young children in Italian slum districts. Like Steiner, she attached great importance to the quality of the early years environment. Furniture was at child height, and storage was designed to be orderly, methodical and accessible. In this way children could begin to take responsibility for selecting activities and for replacing items in their allocated place after use. Again, these are aims which many present-day educators of young children would support wholeheartedly. Montessori believed passionately that young children should use only apparatus made from natural materials of high quality, thus providing pleasurable and stimulating sensory experiences. Teaching apparatus was highly structured and used following highly defined procedures, with the teacher in the role of facilitator: a non-interventionist provider and guide.

This view, however, was not shared by Susan Isaacs (1885–1948), who saw a much more pro-active role for the teacher in maximising the value of play experiences. Isaacs travelled widely from her Cambridge home, where she ran her own school, implementing and developing ideas based on her own careful and methodical observations of children. Lex Grey, a student from New Zealand visiting Isaacs in England during the war, reported, 'This was a woman who understood the children she worked with. This was a woman who actually used to learn from children' (cited in Middleton and May, 1997).

Isaacs recognised the importance of the teacher's own knowledge and its application in devising and presenting activities designed to invite enquiry and exploration. The following paragraph demonstrates how, in many respects, her thinking would seem applicable in spirit to the management of science in a present-day early years environment, although one suspects that current attention to health and safety regulations would preclude several of the implied activities!

> The children are free to explore and experiment with the physical world, the way things are made, the fashion in which they break and burn, the properties of water, gas and electric light, the rain, sunshine, the mud and the frost. They are free to create either by fantasy or imaginative play or by real handling of clay and wood and bricks. The teacher is there to meet this free enquiry and activity by his skill in bringing together the material and the situations which may give children the means of answering their own questions about the world.
>
> (cited in Anning, 1991, pp. 14–15)

Early Years Science in the Statutory Curriculum

During the twentieth century, the education of young children evolved rapidly as methodologies were disseminated, combined and developed to form an amalgam, without specific name, but recognised professionally as 'good early years practice'. Piaget became a household name amongst teachers, and his identification of stages of intellectual development proved pivotal to the changing vision of education. By raising awareness of the child as an active participant in the learning process a greater understanding was born of the need to provide appropriate and challenging classroom experiences.

Jerome Bruner, working in America in the 1960s, was inspired by the work of Russian psychologist Leo Vygotsky (1896–1934). Vygotsky suggested that for any child there is a Zone of Proximal Development. This lies between the level of actual development, which they can achieve without support, and their level of potential development, which can be achieved with the support of a more able 'significant other'. Attempts to develop concepts or skills which are beyond the Zone of Proximal Development for any child would prove futile (see Smith and Cowie, 1991). Establishing existing skills and understandings when teaching science is thus fundamentally important.

Bruner (1966) described a series of progressive ways in which children represent their world. Initially, children enact their experiences through, for example, using play materials. Later, imagery becomes more important as children enter the 'iconic' phase, and it is at this stage that imaginative pursuits such as role play with 'pretending' become most significant. Finally, 'symbolic' representations (principally language) allow children to extend ideas into the abstract. In this way it has been argued that play provides a bridge between concrete experience and abstract thought. If these ideas are accepted, there are far-reaching implications to the teaching of science. If children represent their ideas in these ways, then these are also the routes to learning. Children need to be provided with materials or strong images to 'hold in their heads' as they move towards an ability to adopt abstract concepts. They also need to be provided with opportunities to verbalise their hypotheses and understandings.

Early years science in the 1980s, however, remained generally confined to the contents of the nature table. Her Majesty's Inspectorate's report *The Teaching and Learning of Science*, published in 1989, reviewed developments in primary education over the preceding decade. Despite an encouraging opening statement that 'The amount of science taught in Primary schools has increased markedly over the last ten years and most schools have increased and improved their resources for science teaching', the remainder of the report is less complimentary:

> ... many schools need to improve the breadth and balance of their work
>
> ... a shortage of space in which to carry out practical work
>
> ... difficulties in managing practical activities and group work
>
> ... largely limited to observation and description and mainly to biological topics.
>
> (HMI, 1989)

However, in reviewing examples of good practice in working with 5–7-year-olds, the report finds that 'effective teaching uses carefully selected experiences to build up the children's knowledge and understanding of science' (HMI, 1989). They also cite particular instances of effective skill development. Children were making and recording observations of bubbles, identifying patterns as they explored a range of reflective surfaces, developing hypotheses relating to floating and sinking and 'experimenting' with magnets.

The year 1989 was a watershed for primary education as a whole in that it also marked the publication of the National Curriculum for primary schools. The re-organisation of the primary curriculum was fundamental. As far as science was concerned, the new curriculum was, to say the least, far ranging. The HMI Science Committee concluded that, in addition to the process skills such as 'observing, pattern seeking, explaining, experimenting, communicating and applying' which should be developed through science, the scientific concepts that primary children should meet could be encapsulated in four sections: 'The study of living things and their interaction with the environment', 'Materials and their characteristics', 'Energy and its interaction with materials' and 'Forces and their effects' (HMI Science Committee, 1985). Despite this, the National Curriculum Council managed to produce a science curriculum with seventeen sections, or Attainment Targets. Only three of these were deemed to be inapplicable to children working at Key Stage 1: 'Making new materials', 'Explaining how materials behave' and a cumbersome section which attempted to place scientific methodology in context entitled 'The nature of science'.

The logistical impossibility of planning for the delivery of such a cumbersome and overloaded curriculum in the classrooms of 6- and 7-year-olds was matched only by the confusion caused in the nation's reception classes. The National Curriculum was designed to be taught from Year 1 of the primary school. There was no documentation for reception classes. The statutory nature of the National Curriculum required schools to devise 'maps' which were, in effect, statements of good intentions as they allocated different areas of learning to different year groups. Schemes of work were now referenced to the documentation, showing progression and continuity.

Reception and nursery teachers thus faced a particular dilemma. The general opinion was that, although there was no statutory requirement for them to do so, children in these classes should be carrying out activities relevant to the science curriculum. But it was neither easy nor necessarily appropriate to reference this learning to the Programmes of Study for Key Stage 1. Many head teachers asked their reception teachers to identify, where possible, which aspects of the National Curriculum children were 'working towards' in their tasks. Others, in view of the extent of the curriculum content, decided that the reception children should 'make a start' on it. To counteract the adverse effects of the possible imposition of an inappropriate curriculum on the very young, many schools and local authorities devised their own guidelines for learning in the early years. Not all of these identified science as a discrete area of the curriculum.

Science was designated a core subject in the new National Curriculum. There was now a national entitlement to the coverage of science in primary schools, in many of which it had been either undervalued or almost entirely neglected. A vast in-service training programme raised awareness of how the inclusion of science could enrich the primary curriculum, and gave confidence to those practitioners who felt ill equipped to deliver the subject by rediscovering and developing their personal knowledge. To a great number of primary teachers, the word 'science' merely revived memories which they would have preferred to forget, of their own uncomfortable experiences in secondary school laboratories. How could this subject conceivably be relevant to the early years' curriculum?

Non-statutory guidance, published with the National Curriculum, provided helpful advice, and as teachers began to incorporate science into their planning they developed the skills of identifying opportunities to introduce concepts and skills within the context of their class topics. Popular themes such as 'Ourselves' became standards as rich providers of opportunities to cover aspects of the Attainment Targets, and skills such as those involved in observing different materials were found to adapt to a wealth of diverse contexts.

The PACE project (Primary Assessment Curriculum and Experience) (Pollard et al., 1994) explored the amount of time which teachers of young children were devoting to different aspects of the curriculum and the changes in their identified objectives. Initially they found a marked increase in the amount of time, both in the classroom and in preparation, devoted to science. Marilyn Osborn (1996) interviewed experienced infant teachers and recorded their thoughts about the new orders. One teacher described her difficulties in internalising the new requirements for the teaching of science:

> the science, the approach to science, I mean yes, a lot of it is rooted in good infant practice but it's in a language and a form which is hard for me, and I want to get it into me so that it's happening rather than me thinking, 'Oh, well, I haven't done much on Attainment Target 1, 2b,' rather than me thinking that, that it will be part of my practice.
>
> (pp. 39–40)

Writers such as Wynne Harlen and Sheila Jelly (Harlen and Jelly, 1989) encouraged teachers to appreciate science as a vehicle for many of the important elements of learning in the early years. Through science, for example, children could develop the skills of observation, and from observation would flow understanding and language. By emphasising the role of the teacher in the delivery of the science curriculum, Harlen and Jelly contributed to a constructivist view of teaching, and restored a sense of autonomy to those who had felt undermined by the new curriculum. In describing some of the factors determining children's learning in science, they wrote:

> But what they learn will depend on many things, particularly on what ideas they had at the start, what they did and how they interpreted what they found. In turn, what they do and what they find will depend on the materials that their teacher provided for them to use, the guidance they received and the encouragement to do such things as think things out, check ideas by going back to the objects, improve their technique for finding out, challenge preconceived ideas. For all these things it is the *teacher* who holds the key.
>
> (Harlen and Jelly, 1989, p. 6)

One of the less successful aspects of the 1989 National Curriculum orders for science was that the attempts to place individual concepts on to some sort of hierarchical model within the Attainment Targets were at best contrived and at worst quite ludicrous. At Level 3, for example, which represented the attainment of an able 7-year-old, pupils were expected to know that 'some materials conduct electricity well while others do not', and to understand that 'a complete circuit is necessary for an electrical device, such as a bulb or buzzer, to work'. However, the possibilities of being able to explore either of these aspects would appear to be limited, since it was not until Level 4 that pupils should 'be able to construct simple circuits' (HMI, 1989).

Despite such incongruities, in tandem with the National Curriculum orders for science came the requirement for teachers to assess children's attainment in the subject at the end of Key Stage 1. In 1991, Year 2 teachers suddenly emerged as national trail-blazers as they became the pioneers of this process. They were required to make their own assessments of children's progress, following the guidelines of the Attainment Targets, and to then carry out prescribed tests to further inform themselves, parents and the government.

The ethos of accountability was thus firmly established. Moderators were appointed by local education authorities to oversee the management of the first Standard Assessment Tasks (SATs), and teachers were expected to be able to justify decisions taken in making their own assessments. Although the early SATs in many ways represented excellent practice as teachers recorded and then analysed children's responses to a range of practical activities, it very soon became more than apparent that the administration of these tasks was overwhelmingly time-consuming. The now almost legendary statutory assessment of investigative skills in 1991, for example, involved a small group of children making predictions about the ways in which a group of given objects might behave when placed in a water tank, testing their ideas and recording their results. The complete activity took between forty minutes and an hour. It is anecdotal that the apple which floated at 9.30 in the morning was sinking by 2.00 in the afternoon, as, it is reasonable to suppose, were the spirits of the class teachers!

Monitoring individual progress to inform teacher assessment was no less burdensome. Initially, in the interests of justification, many teachers were required by their local authorities to provide evidence of children's attainment. Some particular authorities became notorious as they demanded examples of children's work which supported every aspect of the science curriculum. In extreme examples, suitcases of paper were amassed by schools and the ultimate repercussions of this were to skew the focus of classroom teaching. The perceived pressure to produce evidence of learning resulted in a high dependence on worksheets and a tendency to ask children to write or draw a record of every experience.

Such pressure was to some extent alleviated by the Dearing Report of 1994, and the publication of the revised National Curriculum orders in the following January. Although the widely publicised 'slimming down' of the curriculum proved to be more an exercise in the removal of repetitious statements across the different subject orders, there was an explicit aim to lighten the load of paperwork for teachers, and the requirement, for example, to keep vast amounts of evidence of individual children's attainment was categorically denied.

The orders for science at Key Stage 1 were by now reduced to four areas, with descriptions of attainment at different levels. There was a fifth area, at the beginning of the orders, which constituted an introductory section to the Programmes of Study. In this section many of the most important aspects of science were to be found. Here were the attitudes and values which may be fostered through science, with references to children relating their new-found understandings to their everyday lives, to an appreciation of the environment and to an awareness of safety issues. It is also in this section that there was explicit reference to children's developing scientific language. Ironically, the lack of a title to this section caused it to be widely overlooked, regardless of the opening statement that its contents were requirements of the Programmes of Study to be applied across the rest of the science curriculum. There was no obvious way of referencing planning to a page with no name.

Sadly, within the remaining sections of the document, Experimental and Investigative Science, Life Processes and Living Things, Materials and Physical Processes, there were some glaring omissions. Despite their acceptance as almost classically suitable areas of exploration for young children, there was no reference to be found, for example, to colour, shadows or seasonal change. Neither does the language of the curriculum denote any particular empathy with the way in which young children learn. There is frequent use of the phrase 'children will be taught', but nowhere is there any hint that they should learn through structured play activities.

Reflections on Current Practice

As an appreciation of the learning opportunities presented by science became more widespread, the subject was evolving into a highly valued component of the early years curriculum. Aware of the importance of placing learning in a context which was relevant and meaningful to young children, teachers became ever more cognisant that science was potentially a highly productive platform for creative and imaginative activities. In nursery and reception units activities were provided which attracted children's attention, evoked inquiry and extended vocabulary in both variety and specificity. Echoes of the 'learning by doing' of Rousseau and Pestalozzi, and the importance of sensory stimulation expounded by Montessori, are still clearly manifest in the science activities of the 1990s.

Froebel, too, would have undoubtedly approved of the careful planning which is now an integral part of the provision of play activities of the early years classroom. With the development of the skills of observation and communication very much in mind, the sand pit and the water tray now undergo regular transformations to become imaginative landscapes such as the 'mini-beast lair' or the 'Arctic water world', enticing children to explore and experience the properties and possibilities of materials and objects. Similarly, a wealth of exploratory scientific play equipment is now routinely seen in the early years, including resources such as kaleidoscopes, colour filters and construction kits, all seen as vital precursors to later learning. In all these activities, learning objectives are explicitly identified and form the focus of gentle adult intervention in the form of questioning and conversation. Froebel would hopefully recognise that this form of play provision is a far from purposeless activity.

Montessori's influence is further demonstrated in the explicit aim of early years educators of enabling children to become independent learners. In addition to the management of classroom resources, mentioned earlier, this has particular relevance to the science curriculum. Many of the children with whom Montessori initially worked came from impoverished backgrounds and were disadvantaged by a variety of learning difficulties. Structure, consistency and procedure were an important source of support in the developing independence of these children. Discussing her didactic materials in *The Advanced Montessori Method* she writes:

> To make the process one of self-education, it is not enough that the stimulus should call forth an activity, it must also direct it. The child should not only persist for a long time in an exercise; he must persist without making mistakes.
>
> (cited in Rusk, 1995, p. 267)

Although the notion of repetition of routine procedures to the point of perfection may not be advocated to quite such extremes in current practice, there are identifiable traces of this methodology in early years science, especially in consideration of investigative processes. The standard, potentially cyclical, experimental procedure of the identification of a point of inquiry, the development of a hypothesis (or the making of a prediction), followed by investigation and finally reflection, is founded in the science experiences of the early years. Primarily, the individual skills of this process are separately identified and encouraged, but the ultimate aim is that the complete process will become internalised through confident familiarity.

Bruner's model of scaffolding as a structure provided by educators to promote and support learning became one of the foundations of classroom practice (Wood et al., 1976). This had particular implications for the development of the investigative process. In the early years, children are developing individual process skills, but are able, within highly structured, teacher-led activities, to carry out the procedure of a simple investigation, exploring a given idea. These early investigations will almost certainly be carried out as groups with intensive adult support. As the investigative framework becomes more familiar, children are able to follow the pattern of an investigation using a prepared template, but it will not be until very much later that children are able to carry out such tasks independently.

Froebel, Montessori, Steiner and Susan Isaacs all believed firmly that the starting point and nature of any new learning is established by the readiness of the child. Isaacs would, one suspects, have found the restrictions imposed by the weight of the modern curriculum on her ability to respond to children's interests and enthusiasms inhibiting to the point of claustrophobia. But there is an appreciation, highly influential in current teaching practice, that children's existing understandings will fundamentally influence their responses to new concepts. Particular attention was drawn to this by the work of various research projects during the last thirty or so years of the twentieth century, notably the SPACE project (Science Processes and Concept Exploration, King's College, London, and Liverpool University). As the role of questioning became ever more prominent as a means of both eliciting information about existing understandings and supporting children in making conceptual connections, so the importance of language development was realised. The science experiences of the early years must generate opportunities for children to extend their vocabularies and develop the skills of communication.

Desirable Outcomes?

One of the main recommendations of the Plowden Report in 1967 had been that nursery education should be available for all children at any time after the beginning of the school year following their third birthday (CACE, 1967). This remains a stated aim of government, yet to be achieved. By the late 1990s, however, strategies were in place to require co-operation between all appropriate providers of day care, including playgroups, nurseries and those reception classes with children under 5 years old, in an attempt to provide potential places for all 4-year-old children, available according to parental choice. In an effort to provide a minimum entitlement of experience, the government published a statutory document which included in its title the wording 'Desirable Outcomes for Children's Learning on Entering Compulsory Education' (DfEE, 1996). The early years curriculum described in this document was divided into six sections: Personal and Social Development, Language and Literacy, Mathematics, Knowledge and Understanding of the World, Creative Development, and Physical Development. It is in the fourth section that the most explicit references to science skills and concepts can be found.

For example:

- *Children* explore and recognise features of living things, objects and events in the natural and made world and look closely at similarities, differences, patterns and change.
- They talk about their observations, sometimes recording them and ask questions to gain information about why things happen and how things work.

However, aspects of the remaining five areas of learning are clearly contributive to science. In Personal and Social Development, children should be 'eager to explore new learning, and show the ability to initiate ideas and to solve simple practical problems', and in Language and Literacy they should 'use a growing vocabulary with increasing fluency to express thoughts and convey meaning to the listener'.

The document thus hopes to lay the foundations of the primary science curriculum: a sense of inquiry, an ability to work collaboratively and a growing capacity to verbalise thoughts and ideas.

Looking to the Future

This integration of curriculum areas, with the exception of literacy and numeracy, seems likely to be a foretaste of the future. Although the possible eventual erosion, albeit implicit, of the position of science as a core element of the Key Stage 1 curriculum may at first sight appear to be a retrograde step, the lack of a heavily prescribed content will conceivably allow teachers more latitude in the choice of contexts they use to develop early skills and processes. In the current curriculum review to be implemented in the year 2000, the 'Desirable Outcomes' will be replaced by 'Early Learning Goals', which will continue until the end of the reception year, thus establishing a Foundation Stage to precede Key Stage 1 (Qualifications and Curriculum Authority, 1999). There is little suggestion of radical change to the tone of the early years science-related curriculum, which remains predominantly embedded within the context of Knowledge and Understanding of the World. At Key Stage 1, the most significant innovation is the higher profile given to attitudes and safety awareness which are embraced in the Programmes of Study. Care for the environment, for example, sensibly becomes integral to a study of living things.

In some ways, science has been ahead of literacy and numeracy. Recent reports from studies of other European approaches to early education indicate that in the UK there has been too great an emphasis on early recording in both these areas. This was explicitly discussed in the final report of the government-appointed Numeracy Task Force, *The Implementation of the National Numeracy Strategy*:

> It is important to stress that the Desirable Learning Outcomes for mathematics should not be interpreted in a way that places too great an emphasis on formal recording of written numerals and calculations. The emphasis, particularly in the very early years, should be to learn about mathematical ideas through counting activities and discussions that enable children to become familiar with numbers and numerals and use these to solve number problems in a practical context.
>
> (Numeracy Task Force, 1998, para. 158, p. 66)

The teaching of science has not, in the main, fallen into this trap, but early years educators must be wary lest it does so. Science must essentially be perceived as a 'hands-on' subject, with a solid appreciation of the fundamental importance of experiential learning. It is in planning, carrying out and evaluating explorations that children's scientific skills are given stimulus and focus. Language development is most certainly a fundamental aspect of early years science, but this must be primarily spoken language. Through verbal communication, describing experiences and developing vocabulary, children will consolidate knowledge, develop and clarify ideas, and make links between new and existing ideas. Imaginative use of role play areas, for example, provides a vehicle for linguistic exploration as children re-enact learning experiences and experiment with ways of using new vocabulary.

An example of this was seen during a recent visit to a reception class. In a role-play kitchen, Jacob was busy making chocolate Rice Krispy cakes. He stirred the imaginary chocolate as it melted in a pan, just as he had seen his teacher do the previous day, and then carefully poured it into a bowl on the table which, he explained, contained the Krispies. As he did so he maintained a running commentary, explaining to me how one has to be careful not to heat the chocolate too much, as it might burn and be spoiled. As I ventured a finger to taste the imaginary mixture I was firmly ticked off. 'Wait for it to cool down!' he remonstrated. After a few moments he decided that the chocolate mix was cool enough to spoon into little cases. These were real paper bun cases, provided as part of the role-play equipment. 'You still can't have one,' he said, in case I was thinking of further tasting. 'The chocolate has to get really cold so it goes hard. I'm going to put them in the fridge. It takes a long time.' At this point the cases were arranged carefully on a small tray and placed in a fridge fashioned from a cardboard box. I explained that I was very unhappy: I had really hoped to enjoy a cake but I had to leave as I had another school to visit. It really was so disappointing. I buried my face in my hands and made every effort to appear devastated. A small face peered closely into mine. 'There aren't really any cakes, you know. It's only pretend.'

In addition, the use of imaginative contexts, including those leading to the production of some artefact, in the presentation of science to young children is not merely a way of 'poaching' additional time for the creative areas of the primary curriculum, laudable as such a motive might be. Science must be perceived as a creative subject in its own right. There is little promise in a future which produces engineers without imagination, or medical pioneers without creativity. By combining the aesthetics of Steiner and the purpose of Froebel, we are able to produce a myriad of contexts within which children can demonstrate understandings and explore applications of their developing knowledge. Too many texts still, for example, show teachers how to develop the skills of electrical circuitry but fail to suggest how they might be used in any meaningful context. How are children to see the value of their skills in such a barren curriculum? How is learning to be perceived as a purposeful occupation?

Recent introductions of the National Numeracy and Literacy Strategies result in a reduction of the time in classrooms available for the rest of the curriculum. Despite its standing as a core subject, science is likely to become increasingly among those subjects described as 'done in the afternoons'. Dadds (1998) uses the phrase 'the hurry along curriculum' to describe a curriculum within which planning for coverage has become a dominant driving force. Science skills are almost certain to be one of the major casualties of this model. Careful observation results from time dedicated to the process, exploring all possible aspects in a thoughtful and unhurried manner. Skills cannot be developed within a pedagogy which is the educational equivalent of a fast food burger bar, speed of delivery being paramount. If pace is the constant objective, children are unable to spend time considering such issues as the most appropriate way of using still unfamiliar vocabulary in phrasing responses, or in making connections between past and new experiences. Dadds (1998) draws attention to the lack of time available for teachers to use in addressing children's individual interests, which are unlikely to fit into the neat compartments of the school curriculum map for their year group. She asks, 'What lessons do children learn about the value and status of their own questions and interests if these are never acknowledged within the curriculum; if they never form the basis for serious learning in schools?'

The way forward must be to develop a broader view of science, particularly in the early years, with a restoration of professional confidence and flexibility in the allocation of time. There is scope for a much greater synergy of the curriculum, with science supplying a valuable springboard for mathematical and linguistic applications, and foundation subjects providing many fruitful contexts for the development of science. There are opportunities waiting to be taken to relate the experiences of science much more closely to generic issues of childhood: environmental awareness, personal safety or the growing awareness of the world of work.

Educators of children in the early years have demonstrated an unrivalled capacity to adapt to change, creatively managing to provide what they believe to be an appropriate curriculum within the constantly shifting constraints of statutory documentation produced in accordance with changing political ideologies. A successful future for science as an element of the early years curriculum is dependent on providing these practitioners with the degree of professional licence necessary to exploit fully their creativity, appreciating the skills, values and attitudes which will be developed.

References

Anning, A. (1991) *The First Years at School. Education 4 to 8*, Milton Keynes: OUP.

Browne, N. (1991) 'The ideological context of science education in the early years: an historical perspective' in N. Browne (ed.) *Science and Technology in the Early Years, An Equal Opportunities Approach*, Milton Keynes: OUP.

Bruner, J.S. (1966) 'On Cognitive Growth' in J.S. Bruner, R.R. Oliver, and P.M. Greenheld (eds) *Studies in Cognitive Growth*, New York: Wiley.

Central Advisory Council for Education (CACE) (1967) *Children and their Primary Schools*, London: HMSO.

Dadds, M. (1998) *Some Politics of Pedagogy*, paper presented at the Association for the Study of Primary Education 'New Minds' conference, St Martin's College, Ambleside, March 1991.

DES (1989) *Science in the National Curriculum*, London: HMSO.

DES (1991) *Science in the National Curriculum*, London: HMSO.

DfEE (1996) *Nursery Education: Desirable Outcomes for Children's Learning on Entering Compulsory Education*, London: DfEE/SCAA.

Frost, J. (1997) *Creativity in Primary Science*, Buckingham: OUP.

Gardner, J. (1993) 'Pressure and the spirit of play', in R. Pusch (ed.) *Waldorf Schools Vol.1: Kindergarten and the Early Grades*, New York: Mercury Press.

Harlen, W. and Jelly, S. (1989) *Developing Science in the Primary Classroom*, Harlow: Oliver and Boyd.

HMI (1989) *The Teaching and Learning of Science, Aspects of Primary Education*, Department of Education and Science, London: HMSO.

HMI Science Committee (1985) 'Science and the Curriculum' in B. Hodgson and E. Scanlon (eds) *Approaching Primary Science*, Harper Education Series, London: Harper and Row.

Middleton, S. and May, H. (1997) *Teachers Talk Teaching 1915–1995: Early Childhood, Schools and Teachers' Colleges*, Palmerston, North Island, New Zealand: Dunmore Press.

Numeracy Task Force (1998) *The Implementation of the National Numeracy Strategy: The Final Report of the Numeracy Task Force*, London: DfEE.

Osborn, M. (1996) 'Teachers Mediating Change: Key Stage 1 Revisited' in P. Croll (ed.) *Teachers, Pupils and Primary Schooling*, New York: Cassell.

Pollard, A., Broadfoot, P., Croll, P., Osborn, M. and Abbot, D. (1994) *Changing English Primary Schools? The Impact of the Education Reform Act at Key Stage One*, London: Cassell.

Pusch, R. (ed.) (1993) *Waldorf Schools Vol. 1: Kindergarten and the Early Grades*, New York: Mercury Press.

Qualifications and Curriculum Authority (1999) *QCA Consults on Early Years Education*, press release, 19 February 1999.

Rousseau, J. (1911) *Emile* (translation B. Foxley), London: Everyman's Library.

Rusk, R.E. (1955) *The Doctrines of the Great Educators*, London: Macmillan.

SCAA (1996) *Desirable Outcomes for Children's Learning*, London: SCAA.

Science Processes and Concept Exploration (SPACE) Project, (1990–3) Liverpool: Liverpool University Press.

Smith, P.K. and Cowie, H. (1991) *Understanding Children's development*, second edition, Oxford: Basil Blackwell.

Steiner, R. (1993) Extract from *The Renewal of Education*, Lecture 5, in R. Pusch (ed.) *Waldorf Schools Vol.1: Kindergarten and the Early Grades*, New York: Mercury Press.
Wood, D.J., Bruner, J.S. and Ross, G. (1976) 'The role of tutoring', *Problem-Solving Journal of Child Psychology and Psychiatry*, 17: 89–100.

4 Learning Concepts

Elaine Wilson

This chapter explores the major theories of how children learn which have been influential in the development of the science curriculum since the early 1960s. The ideas outlined here relate to the parallel initiatives in procedural understanding detailed in Chapter 5. The chapter concludes by considering recent research which challenges the dominant neo-Piagetian perspective held by science educators in the 1990s.

The 1960s saw the first serious attempt at large-scale science curriculum reform, culminating in the Nuffield 5–13 science courses. The intention of this development was that pupils would be encouraged to discover science for themselves and focus on scientific methodology rather than scientific 'facts'. Behind this was the underlying assumption that pupils did not have any ideas already about scientific concepts, and that they would discover the laws of science through observation and practical activities. In essence, the Nuffield project was 'to awaken the spirit of investigation and to develop disciplined imaginative thinking' (Nuffield Foundation, 1966). The Piagetian theories advocated by the Plowden Report were influential in writing the scheme, and it is this recognition of the importance of cognitive processes that was innovative and underpinned the Nuffield philosophy. However, the resultant 'recipe style' Nuffield package proved to be a distortion of the discovery model of the 'pupil as scientist' in action. Each major topic included in the package had an accompanying set of prescriptive pupil books and teacher's guides which led every pupil along the same line of thought. The activities were often so tightly controlled that the 'right answer' was apparent to pupils. The net effect was that there was little opportunity for discovery learning, and lessons became simply illustrations of key concepts with little or no imaginative thinking taking place. The guided discovery ideal became, in many schools, stage-managed heurism, and many science educators questioned the value of such an approach. The Nuffield approach failed, too, to take account of the learner's prior knowledge.

The Nuffield secondary science schemes had been designed with the most able grammar school pupils in mind. The primary schemes were not widely adopted, as science did not have the same status in the primary curriculum. The introduction of National Curriculum requirements has only recently ensured that children in this phase of education have the opportunity to access a broad and balanced science education. As the nature of the education system changed and moved towards comprehensive education, it became clear that the Nuffield approach was not appropriate for the vast majority of pupils in secondary schools. The general dissatisfaction with the Nuffield schemes among science education researchers, alongside the demands made on teachers in comprehensive schools to cater for a full ability range intake, initiated a search for an alternative approach (Cheung and Taylor, 1991). The science educators and curriculum developers of the time looked to other emerging theories for an alternative teaching methodology. The constructivist ideas which emerged, and which are discussed later, then became the 'generally accepted philosophy stream' (Nussbaum and Novick, 1982). However, thirty years later constructivism seems to have reached a plateau and science educators are looking beyond constructivism. Why has constructivism been so influential, and why now does it appear to be falling out of favour? Before fully considering this question, I will digress slightly and consider both a definition of a concept and recent research into conceptual change. This is necessary as constructivist theories are based on notions of conceptual development.

Concepts and Conceptual Change

Concepts allow us to classify and process incoming information by drawing on past experiences. Successful integration of new concepts takes place when the learner is able to categorise and store ideas in order to deal with them more efficiently. Conceptual change happens when naïve beliefs about natural and social phenomena are replaced by more sophisticated ones. The belief that this is an important process in science education is now widely accepted.

In the early 1970s research in science education began to focus on the conceptual models that lie behind reasoning patterns used, particularly in the physical sciences. Researchers went into science classrooms and saw for themselves what children's experience of science education was really like. The Assessment of Performance Unit (APU), referred to throughout this book, was set up in 1975 and provided the forum for much of the classroom-based research. This DES-funded project was intended to produce a national picture of pupil performance. The survey teams used one-to-one interview, discussion based on drawings, and written questions to collect data. Children at 11, 13 and 15 years (Eggleston, 1991) were surveyed in schools throughout the country and the result was an extensive database of pupil answers to questions about scientific concepts. In 1978 the Primary Survey reported that few primary schools offered systematic teaching of science concepts and that, in fact, over a fifth of schools offered no science at all (DES, 1978). They commented too on the generally poor match between the standard of work achieved and the level children were thought capable of reaching. The worst matches were found in science, geography and history, with match being defined by Harlen (1992) as

> finding out what children can already do and what ideas they have, as a basis for providing experiences which develop these skills and concepts.

> The keynote of matching is this finding the right challenge for a child, the size of the step that he can take by using but also extending existing ideas.
> (p. 69)

(See also Chapters 2, 5 and 6 in this volume.)

In 1985 the Nuffield Foundation agreed to fund a project that was concerned with the investigation of the ideas that primary school children (5–11) had about the scientific aspects of the world around them, and with finding ways of encouraging the development of the children's ideas towards the more widely applicable scientific view. The Science Processes and Concepts Exploration (SPACE Research Reports 1990–91) project went further and attempted to identify the scientific ideas that are relevant at pre-secondary level and to map out the development of these ideas. These surveys collected vast amounts of information which were to inform subsequent curriculum development. The APU and SPACE framework was based on a view that there is a common scientific methodology based on skills, processes and concepts. The influence of this structure is very evident in the school science curriculum review which was initiated in the 1980s in the midst of this torrent of work. The key players at the time were the Nuffield Primary Science Project, the Secondary Science Curriculum Review (SSCR) and the Children's Learning in Science (CLIS) Project. The CLIS project, particularly, drew on constructivist theories of learning .

What is Constructivism?

Constructivism is grounded in cognitive psychology. Constructivist theory in education is a branch of neo–Piagetian thought which is rooted in Personal Constructivism (Novak, 1977; Von Glaserfeld, 1989). Solomon (1987), Millar (1989a), and Cobern (1993) have taken Personal Constructivism further and have paved the way for Social Constructivism, which is defined by how the learner interprets phenomena and internalises these interpretations in terms of their previous experience and cultures.

The work of Piaget had informed the Plowden Report in the 1960s and was again drawn on in the 1980s. As Vygotskian ideas became more widely known, they too were woven into the social constructivist model, in response to the acceptance that social processes had a vital part to play in knowledge construction, both at the level of the individual (Edwards and Mercer, 1987; Solomon, 1987) and within the community of scientists. Both Piaget and Vygotsky held that

- each individual constructs his/her own knowledge and meaning;
- children's thinking is constrained, and high-level intellectual functions are not available until adolescence.

Table 4.1 compares the theories of Piaget and Vygotsky.

Table 4.1: A comparison of the theories of Piaget and Vygotsky

	Piaget	*Vygotsky*
	an epistemologist seeking the origins of knowledge	a psychologist seeking the origins of consciousness
Concepts	spontaneous and non-spontaneous	arise from child's independent thinking
Learning	an internal process intrinsically linked to the stage of maturation	an external process in direct response to social world in which the child is surrounded
Language	related to the stage of development and subordinate to development	an on-going process which develops in response to contact with teachers and everyday experiences

Learning science was believed to involve more than the individual making sense of their personal experiences and to be more about being initiated into the 'ways of seeing' which have been established and found to be helpful models by the scientific community. Social constructivists argued that such 'ways of seeing' could not be discovered by the learner and believed that, in the unlikely event of them stumbling upon the consensual viewpoint of the scientific community, the learner would be unaware of the status of the ideas.

Towards a Theory of Learning in Science

The relationship between theory and practice is a symbiotic one: theory contributes to practice, and practice contributes to theory. Social constructivist views were endorsed by the CLIS and SPACE curriculum projects. The ultimate aim of both projects was to address the difficulties of abstract concepts and technical language highlighted by the APU and SPACE surveys. The science curriculum was viewed by the CLIS team as a set of experiences from which the learners construct a view closer to the scientific view. The role of the teacher is as a mediator between scientists' knowledge and children's understanding, acting as a 'diagnostician of children's learning' (CLIS, 1987), at the same time planning out a route to enable conceptual development.

The CLIS (1987) project was less interested in how many pupils got the right answers to the APU survey questions and more in the quality of the 'wrong' answers given. This strand of their research work focused on what have been called 'misconceptions, preconceptions, alternative conceptions', or 'children's science', depending on the philosophical views of the research teams involved. The most recent account of this is presented in Driver and Easley (1978), where the phrase 'alternative frameworks' first appeared. This is now the most widely held definition of the idea that children already have a human-centred view of the natural world which is anchored in everyday experience.

Nearly thirty years after Nuffield 5–13, there is an extensive literature indicating that children come to their science classes with prior conceptions that may differ substantially from the ideas to be taught, that these conceptions influence further learning, and that they may be resistant to change. Despite being geographically dispersed, this work clearly forms a distinctive research programme which Millar (1989a) describes as the 'Alternative Conceptions Movement'. There are now many papers written concerning children's ideas about the whole range of concepts encountered by pupils in the National Curriculum (Brook et al., 1984; Driver et al., 1994; Solomon, 1983).

Introducing the Models into Science Classrooms

The two most influential constructivist models of instruction have been CLIS and SPACE (see Table 4.2).

The instructional stages broadly parallel those of the learning model: elicitation of prior ideas, their clarification and change within the class group, exposure to conflict situations and reconstruction of naïve ideas to more scientifically accepted ones. Both CLIS and SPACE projects are concerned with the range and variety of teacher tasks involved and are considerably more complex than simply 'telling pupils the right answer'. The approach asks teachers to adopt a series of roles in the classroom which are difficult to undertake, and this may have been why they have not been universally adopted. Adapted from Watts (1992), some of the teacher roles are given below:

learner	learning to think like a pupil
psychologist	making diagnoses and prognoses throughout lessons
epistemologist	exploring the nature of knowledge and evidence
practical philosopher	testing the nature of meanings in the classroom
author	developing ideas and notions
field researcher	constantly monitoring data and effect
natural philosopher	being a scientist too

The Cognitive Acceleration though Science Education (CASE) project appeared in the mid 1980s and was another method of instruction based on neo-Piagetian models. The project was developed by Adey et al. (1989) from

Table 4.2: CLIS and SPACE models of instruction

SPACE	*CLIS*
Providing opportunity for exploration and involvement	
Finding out current ideas	Orientation: eliciting ideas
Reflecting on experiences which have led to currently held views	Restructuring of ideas, involving clarification of exchange, exposure to conflict situations, construction and evaluation of new ideas
Helping children to develop their ideas and process skills	Application of ideas
Assessing change in ideas and process skills	Review of change in ideas

King's College, London, and was an intervention programme stemming from Piagetian and Vygotskian psychology. CASE was intended for use in the early years of secondary education to coincide with the most rapid Piagetian maturation stages. The CASE activities were developed following a study of research on Cognitive Acceleration, and the curriculum materials produced were an attempt to translate the theoretical ideas into activities which were of practical use in schools. The central tenet of CASE is that a particular set of teaching strategies can accelerate children's intellectual development and, in the longer term, procure enhanced academic achievement. In essence, the strategies begin by setting a scientific concept in context: concrete preparation. This is followed by presenting the child with a situation in which they are forced to reassess their own ideas: cognitive conflict. At this point the teacher intervenes to enable learning or metacognition to take place. The cognitive conflict and metacognition steps are encompassed in a hypothetical conceptual construction zone labelled the Zone of Proximal Development (ZPD) by Vygotsky. Each child, he suggested, has a personal ZPD that represents 'the distance between the actual development as determined by independent problem-solving, and the level of potential development as determined through problem-solving under adult guidance' (Vygotsky, 1978). The extra amount of constructing that can be done by the child depends on the child's current level of development, what the child already knows, and the discourse between child and environment.

It is this process, Adey (1992) asserts, in which the teacher bridges the gap that enables the pupil to learn. This intervention has been likened to the construction of scaffolding by Bruner (1985). The CASE intervention is designed to accelerate development so that pupils progress from Piagetian 'concrete thinking' to 'formal operational thinking' by the end of the two-year programme. The number of secondary schools using CASE materials has risen in the last ten years and coincides with the introduction of league tables and the comparison of a school's percentage A–C grades as a measure of success. The CASE materials suggest that pupils in the GCSE D-grade category are most likely to benefit and improve their performance as a result of direct intervention at Key Stage 3.

Why Are We Now Looking Beyond Constructivism?

The usefulness of the social constructivists' methodology is now being questioned (Millar, 1989a; Osborne, 1996). Millar challenges the need for a constructivist method of instruction and argues that if such a model of learning is valid, then some learners are able to internalise concepts without such overt teaching strategies. Otherwise, he argues, it is difficult to account for the fact that many people who have not had constructivist instruction 'understand' scientific concepts. It may well be, then, that there are other factors at work in learning scientific concepts.

Recent research (Watts and Alsop, 1997) on motivation linked to learning suggests that pupils are unlikely to be willing to take on the serious intellectual work of reconstructing scientific meaning unless they can see the point of what they are doing. Many of the questions pupils are asked in science lessons, particularly in secondary school, are described by Johnstone (1991) as 'so what' questions. Secondary pupils are no longer curious, on the whole, as to why grass is green or blood is red. Formal school science is about looking for large, long-ranging theories and hypotheses to explain systems like, for example, the laws of motion, the behaviour of tides or the nature of matter. This type of science reasoning quite often does not follow common-sense hunches (Wolpert, 1992) but can be very awe-inspiring. However, its significance is lost on many young people who are living perfectly well despite their ignorance of the bigger picture. These theories, particularly in physical science, are often to do with ideas anchored in abstract reasoning. Millar suggests that constructivist instruction may not lend itself to some of these concepts:

> Constructivist science educators acknowledged that science as a school subject poses formidable challenges to the teacher in maintaining the involvement of many pupils simply because the science covered at school is almost entirely a consensually agreed body of knowledge. There is, therefore, a limited value in children taking away from science lessons ideas that diverge radically from the accepted ones.
>
> (Millar, 1989a, p. 590)

Promoting conceptual change means, too, that the science teacher must ensure that the pupil is presented with a more plausible model than the one already held if the accepted ideas are to be successful in replacing existing ideas. For example, pupils are taught that plants make wood from a reaction involving carbon dioxide gas in the air. Many, however, persist with the idea that wood is produced from soil being sucked up into the stem of the plant, despite being taught about photosynthesis (Brook et al., 1984). This soil idea seems more intuitively accurate than the explanation that solid wood has been made from a gas. Being made aware of commonly held alternative frameworks could assist the teacher in promoting conceptual change. Whilst there may be justifiable reservations about this approach, there is now a readily available source of data through the SPACE and CLIS projects which could inform planning. The danger is that teachers adopt commercial schemes which save time and effort, but in the process lose sight of the research which suggests better progression in learning key concepts.

Secondary science teachers, particularly, have been plagued by externally imposed innovations for over ten years, and this has driven many teachers to resort to lessons which are dominated by the transmission of 'facts' from the teacher to the pupil's exercise book. The net result is that the knowledge cannot be challenged by the learner and there is little scope for any creative involvement. The everyday language used outside science lessons clashes, too, with precise scientific definition. Teachers often fail to allow pupils to engage in talking about their ideas in science lessons (Wilson, 1999). Shallow rote-learning is encouraged, and more meaningful learning of concepts does not take place (Entwistle and Ramsden, 1983; Johnstone, 1991).

Schools have looked to CASE as 'a sort of diet supplement that is not part of science but renders that science more intelligible' (Jones and Gott, 1998) so that pupils will be able to cope with difficult scientific concepts through the development of 'high order thinking skills'. However, critics of CASE (Leo and Galloway, 1996) argue that the programme's success can be accounted for by simple changes to teaching strategies. A key element of CASE lessons is the requirement to run both a co-operative and an individualistic classroom environment. In some cases, the arrangements of children into small groups for work is often a deviation from the normal whole-class situation for both the teachers and children. It has been reported that pupils working in small groups are able to manage their own time and activity more efficiently (Mecce et al., 1988), and that interest in the subject, feelings of competence and problem-solving skills are also likely to be enhanced (Ames and Ames, 1984). So it is not unreasonable to suggest that some of the success of CASE is due to a change in lesson structure which might have inspired positive changes in children's learning goals and motivation by giving them more control over their own learning. Leo and Galloway (1996) have suggested that children's awareness of and control over their own thinking and learning processes is linked to metacognition.

Where Are We Now?

The National Curriculum has ensured that science is now an important part of the learning for all children 5–16. More pupils than ever before follow a 'broad and balanced' education between these ages. Both national and international indicators point to the success of these changes, particularly at the primary stage. Ofsted reports have judged 80 per cent of primary school science lessons to be satisfactory or better (Ofsted, 1997). The Third International Mathematics and Science Survey (TIMSS: Harmon et al., 1997) has shown improvements in performance in science. However, despite science now being a core subject for pupils between the ages of 11 and 16, fewer pupils continue with science beyond 16. Pupils frequently spend a lot of time in Year 7 repeating concepts already studied at Key Stage 2 and complain about science being boring at Key Stage 3. These issues have been noted and considered by the *Beyond 2000* report (Millar and Osborne, 1999) which was funded by the Nuffield Foundation (see Chapters 9 and 11).

Where Next?

Perhaps science educators need now to look beyond the confines of cognitive psychology in developing pupils' understanding of scientific concepts. They might consider the processing models favoured by many mainstream psychologists (Greene and Hicks, 1984) and already adapted for learning chemistry at undergraduate level by Johnstone (1991). (See Figure 4.1.)

Figure 4.1: A processing model of learning

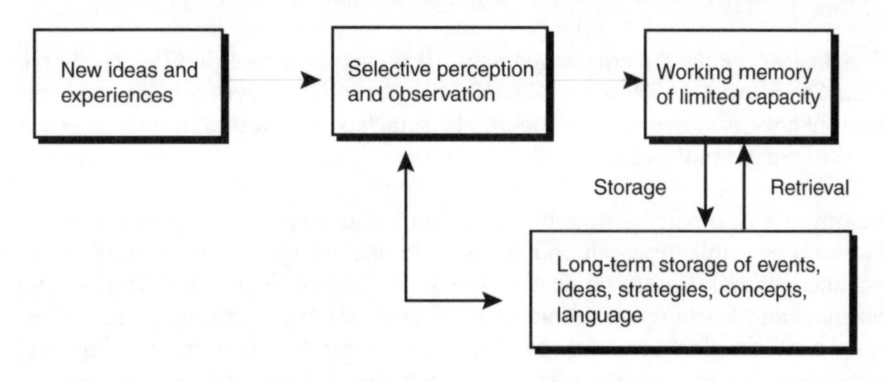

Source: Adapted from Johnstone (1991)

However, there are four immediately accessible points that practising teachers might consider in teaching concepts to pupils between the ages of 3 and 13.

1 Recognising what pupils already know The SPACE and CLIS projects have assembled detailed information about children's ideas or 'alternative frameworks'. This information could be made more widely available, perhaps through the non-statutory guidance for the National Curriculum. Teachers could be trained to use techniques other than pencil-and-paper tests to diagnose what pupils know already. Concept maps and concept cartoons are now used successfully at Key Stages 1 and 2 (see Chapter 6).

2 Teach fewer concepts The curriculum changes which took place in the late 1980s and early 1990s were not accompanied by a corresponding change in the content, particularly at secondary level. The new statutory orders for science have done little to reduce the content and so allow teachers time to explore and consolidate concepts. Sadly, the NC continues to encourage teachers to adopt a 'coverage of the curriculum' mentality, with little opportunity for pupils to question.

3 Improve continuity across key stages and progression of the development of concepts. Pupils are exposed to scientific concepts at a much earlier stage in their education The Piagetian ideas of stages of maturation of the child and the fear that pupils would not be ready to learn some of the abstract concepts have not proved to be the case. Pupils are emerging from Key Stage 2 with well-formed ideas. This is not often recognised by teachers at Key Stage 3. The 'alternative conception movement' could provide a framework for the progression of concepts and suggest a sequence for learning these.

4 Acknowledge the diversity of learners It is clear now that pupil motivation is linked to meaningful learning. Encouraging secondary teachers in particular to explore a wider range of teaching and learning styles to respond to the needs of all the learners in the class would go some way to helping.

Learning is not the goal in many classrooms. The emphasis is on time on task. However sensible research might seem to be, its effectiveness in terms of enhanced pupil classroom learning will be limited if the implications for teaching and learning are not adopted by teachers. It is vital that there is effective dissemination of the products of research through training and professional development programmes. However, if the divide between research evidence and current pedagogical practice is to be bridged, there needs to be more direct involvement of teachers in the research programmes.

References

Adey, P. (1992) 'The CASE results: implications for science teaching', *International Journal of Science Education*, 14 (2): 137–46.

Adey, P., Shayer, M. and Yates, C. (1989) *Thinking Science: The Materials of the CASE Project*, London: Nelson.

Ames, R. and Ames, C. (1984) *Research on Motivation in Education 1: Student Motivation*, London: Academic Press.

Brook, A., Briggs, H., Bell, B. and Driver, R. (1984) *Aspects of Secondary Students' Understanding of Plant Nutrition: Summary Report*, Leeds: CLIS Project.

Bruner, J. (1985) 'Vygotsky: a historical and conceptual perspective' in J. Wertsch (ed.) *Culture, Communication and Cognition: Vygotskian Perspectives*, Cambridge: Cambridge University Press.

Cheung, K. and Taylor, R. (1991) 'Towards a humanistic constructivist model of science learning: changing perspectives and research implications', *Journal of Curriculum Studies*, 23 (1): 21–40.

CLIS (1987) *Children's Learning in Science Project*, Centre for Studies in Science and Mathematical Education: University of Leeds.

Cobern, W. (1993) 'Contextual constructivism' in K. Tobin (ed.) *The Practice of Constructivism in Science Education*, Washington DC: AAAS, pp. 51–69.

DES (1978) *Primary Education in England: A Survey by HM Inspectors of Schools*, London: HMSO.

Driver, R. (1989) 'Students' conceptions and the learning of science', *International Journal of Science Education*, 11 (Special Issue): 481–90.

Driver, R. and Easley, J. (1978) 'Pupils and paradigms: a review of literature of concept development in adolescent science studies', *Studies in Science Education*, 5: 61–84.

Driver, R., Squire, A., Rushworth, P. and Wood Robinson, V. (1994) *Making Sense of Secondary Science*, London: Routledge.

Edwards, D. and Mercer, N. (1987) *Common Knowledge: the Development of Understanding in the Classroom*, London: Methuen.

Eggleston, J. (1991) *The 1980–82 APU Science Survey. An Independent Appraisal of the Findings*, London: School Examination and Assessment Council (SEAC).
Entwistle, N. and Ramsden, P. (1983) *Understanding Students' Learning*, London: Croom Helm.
Greene, J. and Hicks, C. (1984) *Basic Cognitive Processes*, Milton Keynes: Open University Press.
Harlen, W. (1992) *The Teaching of Science*, London: David Fulton.
Harmon, M., Smith, T.A., Martin, M.O., Kelly, D.L., Beaton, A.E., Mullis I.V.S., Gonzalez, E.J. and Orpwood, G. (1997) *Performance Assessment in IEA's Third International Mathematics and Science Study (TIMSS)*, Boston: Center for the Study of Testing, Evaluation and Educational Policy.
Johnstone, A. (1991) 'Why is science difficult to learn? Things are seldom what they seem', *Journal of Computer Assisted Learning*, 7: 75–83.
Jones, M. and Gott, R. (1998) 'Cognitive acceleration through science education: alternative perspectives', *International Journal of Science Education*, 20 (7): 755–68.
Leo, E. and Galloway, D. (1996) 'Conceptual links between cognitive acceleration through science education and motivational style: a critique of Adey and Shayer', *International Journal of Science Education*, 18 (1): 35–49.
Mecce, J., Blemfield, P. and Hoyle, R. (1988) 'Students' goal orientation and cognitive engagement in classroom activities', *Journal of Educational Psychology*, 80 (4): 514–23.
Millar, R. (1989a) 'Constructive criticisms', *International Journal of Science Education*, 11 (Special Issue): 83–94.
Millar, R. (ed.) (1989b) *Doing Science: Images of Science in Science Education*, London: Falmer Press.
Millar, R. and Osborne, J. (1999) *Beyond 2000: Science Education for the Future*, King's College London. Available on http://www.kcl.ac.uk/education
Novak, J. (1977) *A Theory of Education*, Ithaca NY: Cornell University Press.
Nussbaum, J. and Novick, S. (1982) 'Alternative framework, conceptual conflict and accommodation', *Instructional Science*, 11: 183–208.
Ofsted (1997) *Standards in Primary Science*, London: HMSO.
Osborne, J. (1996) 'Beyond constructivism', *Science Education*, 80 (1): 53–82.
Shayer, M. (1996) *The Long Term Effects of Cognitive Acceleration on Pupils' School Achievement*, November 1996. Available on http://www.thenerve2com/ca/NewsRes.html
Solomon, J. (1983) 'Learning about energy: how pupils think in two domains', *European Journal of Science Education*, 5 (1): 49–59.
Solomon, J. (1987) 'Social influences on the construction of pupil's understanding of science', *Studies in Science Education*, 14: 63–82.
Solomon, J. (1995) 'Constructivism and quality' in P. Murphy, M. Selinger, J. Bourne and M. Brigg (eds) *Subject learning in the Primary Curriculum*, London: Routledge.
SPACE (1990–94) *Research Reports*, Liverpool: Liverpool University.
Von Glaserfeld, E. (1989) 'Cognition, construction of knowledge and teaching', *Synthese*, 80 (1): 121–40.
Vygotsky, L. (1978) *Mind in Society*, Cambridge, Mass.: Harvard University Press.
Watts, M. (1992) 'Children's learning of difficult concepts in chemistry' in M. Atlay, S. Bennett, S. Dutch, R. Levinson, P. Taylor and D. West (eds) *Open Chemistry*, London: Open University.
Watts, M. and Alsop, S. (1997) 'A feeling for learning: modelling affective learning in school science', *Curriculum Journal*, 8 (3): 351–65.

Wilson, E. (1999) 'The use of language in science' in E. Bearne (ed.) *The Use of Language Across the Secondary Curriculum*, London: Routledge.
Wolpert, L. (1992) *The Unnatural Nature of Science*, London: Faber and Faber.
Woolnough, B. (1994) 'Factors affecting students' choice of science and engineering', *International Journal of Science Education*, 16 (6): 659–76.

5 Developing a Scientific Way of Working with Younger Children

Paul Warwick

In this chapter I will be exploring the ways in which the processes and skills that make up the scientific way of working have been given emphasis by different bodies and in different ways since the 1960s, and the effect that this has had on teaching and learning, particularly in the primary school. I will examine the ways in which the development and structure of the National Curriculum has led to particular emphases with regard to the scientific way of working and on associated practice in the classroom. Finally, I will give some attention to the extent to which teachers seem to understand why the way of working in science is important. This, together with an examination of government curriculum initiatives, will lead to grounded speculation about the future. It will, therefore, be clear that this chapter has strong links with Chapters 2, 4 and 6.

It is helpful at the outset to come to some definition of the scientific way of working and its relationship to the processes and skills of science. Prominent among those who have considered this has been Wynne Harlen, whose numerous publications, research reports and evaluations have furnished us with a clear understanding of the important components of scientific investigation (see particularly Harlen, 1996, pp. 26–34, and Russell and Harlen, 1990, for the STAR research project). For the purposes of this chapter it is useful to think of the scientific way of working as an over-arching idea – an approach to working that all scientists have internalised and are able to apply in new contexts. It might be thought of as being constructed of a few central *processes* – hypothesising, devising an investigation, carrying out the investigation, recording, interpreting and communicating – each of which is underpinned by essential practical and intellectual *skills*. These skills are developmental and include observation, measurement, manipulation of variables, prediction, tabulation and graphing, raising questions, using secondary sources of information, and the manipulation of equipment. Though it will become clear in the later discussion of ideas about procedural understanding that this is a simplistic model, it provides a basis for considering the attention that this 'side of science' has received in the curriculum of the 3–13 age range since the early 1960s.

The Centrality of Process Skills in Curriculum Innovation and Debate

The Primary Survey of 1978 provides a useful starting point for setting current thinking about the scientific way of working, and speculation about the future for work with younger children, in context. In it, HMI stated:

> Few primary schools visited in the course of this survey had effective programmes for the teaching of science. There was a lack of appropriate equipment; insufficient attention was given to ensuring proper coverage of key scientific notions; the teaching of processes and skills such as observing, the formulating of hypotheses, experimenting and recording was often superficial. The work in observational and experimental science was less well matched to children's capabilities than work in any other area of the curriculum.
>
> (DES, 1978, para. 5.66)

The picture painted is one of little interest in promoting the subject at primary level and little systematic teaching that might help in the building of key concepts or in the development of the processes and skills necessary for the structured collection of evidence. Within this picture, the movement to secondary school might be characterised as the first time that pupils were able to encounter a structured body of science knowledge and the first time that science processes and skills were given prominence. The scientific way of working was, by this account, largely a mystery to primary pupils.

Certainly there was further evidence to support this view in analyses of the take-up of the major curriculum science projects of the previous ten years or more (Steadman, 1978). An alternative view, however, might be that this largely ignores the debates and the associated curriculum developments that, from the early 1960s onwards, had brought the very notion of science in the primary school into the spotlight, and which were slowly, if perhaps too slowly, making inroads into the landscape of the primary curriculum. Progress through the 1960s and 1970s *was* partial, so did the emphases given to curriculum developments and to the debate about what should be taught from 5 to 13 actually hold back the progress of science teaching? Or did they lay the foundation for an appropriate emphasis on process skills in later developments and legislation? Certainly, nearly all of the developments of this period were very firmly based upon the belief that what children should encounter was work that emphasised science as a way of working rather than science as a particular body of knowledge. It is therefore worth spending some time examining these developments before returning to the question of their contribution both to the science curriculum for younger children and, as a consequence, to children's understanding of the scientific way of working.

The 1960s saw an unprecedented interest in the science curriculum of younger children, largely brought about by the work of the Schools Council and Nuffield Foundation projects. And other forces were also at work. In 1961, school inspectors praised the work of teachers who were attempting more than nature study in this curriculum area (Ministry of Education, 1961), while in 1963 the Association for Science Education set up a Primary Schools Science Committee. From the outset the emphasis was clear, with the Committee making a number of assertions in their policy statement reflecting their belief that the teaching of a body of knowledge was of relatively little importance compared to the development of enquiring young minds (ASE, 1963). Taking a lead from such statements, the development of curriculum science materials, and the associated research that had led to this development, placed science process skills at the heart of what should be emphasised in both the primary and lower secondary science curriculum. The first project of interest was the Nuffield Junior Science Project (1964–6), which expressed a strong belief in the idea that children's own ways of solving problems were essentially scientific and that these problem-solving capacities could be enhanced through following children's leads appropriately.

> Children's practical problem solving is essentially a scientific way of working, so the task in school is not one of teaching science to children, but rather of utilising the children's own scientific way of working as a potent educational tool.
>
> (cited in Black, 1980)

The project books gave advice on the ways in which children's ideas might be followed and how this might lead to fruitful scientific endeavour, but little attempt was made to place this advice within a framework of content beyond providing some starting points for study. Guided by informed teaching, the content was to be primarily the problems that children found to be of significance to themselves, while 'the greater emphasis ... [was] on the so-called processes – observing, pattern-seeking, hypothesising and planning experiments' (Wastnedge, 1983).

At roughly the same time, the Oxford Primary Science Project, with Ministry of Education funds from 1963 to 1967, was focusing more directly on work whose aim was the exploration and development of children's science concepts. It is interesting to note that, in many ways, Nuffield Junior Science and the Oxford Primary Science Project were mirrored by similar developments in the United States. There, the Science Curriculum Improvement Study (1970) tended to place *what* children were to learn before *how* they were to learn, while the Elementary Science Study that ran from 1960 gave rather more weight to the development of scientific enquiry.

The immediate successor to the work of the Nuffield Junior Science Project was Science 5–13 (1967–74), jointly funded by the Schools Council, the Nuffield Foundation and the Scottish Education Department. This shared many of the educational convictions of the Nuffield Junior Science Project. At a time when Plowden (Central Advisory Council for Education, 1967) was placing a stress on the value of individual discovery and first-hand experience, Science 5–13 was in keeping with the spirit of the age. The stress was to be on discovery learning, but planned and guided discovery. For this, teachers' materials were produced that provided a set of over 150 behavioural objectives, strongly linked to Piagetian developmental stages, to guide opportunities for learning. In addition, project units gave suggestions for experiments or activities that might be linked to topics on which the children were engaged. A watered-down conventional secondary school science curriculum was unequivocally rejected. Rather, the project had three main areas of concern: the provision of structured learning objectives to be pursued by teachers, strongly linked to the development of the science process skills fundamental to the scientific way of working; providing suggestions for appropriate activities that might link to topics in which the children could be actively engaged in solving their own problems; finally, the provision of a range of objectives that stressed the links between science and other areas of the curriculum. The emphasis was on children learning with understanding and on acquiring the skills of scientific enquiry that would allow them to solve their own problems.

Following from Science 5–13, the Schools Council fostered Progress in Learning Science (see Harlen et al., 1977). Once again, the focus was on observation, the perception of patterns, the formation of hypotheses and the design of experiments, but this project concentrated on the appraisal of individual progress through a scheme based on the developmental objectives of Science 5–13. As the 1970s ended, the Learning Through Science Project, again sponsored by the Schools Council, was well under way. The development of pupil materials was to be fundamental. Teacher groups identified key themes such as Ourselves, Sky and Space, Colour and Materials, around which activities could be written. But it would be a mistake to characterise this project as content-led. Fundamental to decisions made about content was the requirement that the work should match Stage 2 of the developmental sequence set out by the Science 5–13 project. The requirement to focus on science processes and the need for children to be actively involved in their own scientific exploration of the world were both very much in evidence in the thinking behind the project, which was to reach full fruition in the mid 1980s.

Into the 1980s – Nothing Achieved?

It is clear that by the start of the 1980s several large-scale projects, some with significant associated curriculum materials, had attempted to move science teaching and learning forward, particularly within the primary sector. Most placed the development of a scientifically enquiring mind, with the associated development of the processes and skills of the scientific way of working, at their heart. Yet the evidence of the Primary Survey (DES, 1978) and the virtual despair of some influential educationalists at the start of the decade (see Black, 1980) suggests that very little had been achieved in the previous twenty years. For such commentators, the exhortations of Prime Minister James Callaghan's Ruskin speech for schools to 'equip children to the best of their ability for a lively, constructive place in society' (quoted in Brooks, 1991) had borne little fruit with respect to science in the primary school. The worst fears of Professor

Bantock in the first of the Black Papers seemed to such individuals to have a particular potency when applied to the science curriculum: 'at primary school some teachers are taking to the extreme the belief that children should not be told anything, but must find out for themselves' (quoted in Brooks, 1991).

Certainly, Steadman (1978) reports that only 18 per cent of a random sample of schools were even familiar with the Nuffield Junior Project, with 30 per cent of the same sample being familiar with the Science 5–13 project. The Primary Survey refers to the almost negligible impact of these and other curriculum development projects, stating that despite them, 'the progress of science teaching in primary schools has been disappointing' (DES, 1978).

Several reasons are stated for this apparent lack of progress. Fundamental ones were: a lack of understanding among head teachers of the important contribution that science can make to children's intellectual development; a lack of working knowledge among teachers of what constitutes appropriate science for primary children; and a lack of appropriate training of teachers, both initial and in-service, to encourage the effective teaching of science. There were undoubtedly pockets of 'exciting and worthwhile practice' but the problem was 'getting all to do it!' (Richards, 1985).

Reflecting on these problems in the light of the work of the various curriculum projects that have been discussed, the question of the scientific literacy of teachers and head teachers is clearly of vital importance. In placing scientific skills and processes at the centre of developments, and in avoiding the possible 'easy option' of providing a set of clearly structured content-led activities for children to work through in favour of a concern with the development of scientifically enquiring minds, perhaps the task of implementation was made too daunting for teachers with a limited understanding of science who wished to set out on the road of implementing worthwhile science in their schools. Numerous schools acquired, in particular, the excellent Science 5–13 teacher materials, yet the reaction to them in many cases was perhaps not what might have been anticipated by their originators. The objectives were often (though not, it must be stressed, always) either disregarded or, looked at through post-Plowden eyes, seen as something that the school was doing already in all its work. This last reaction was partly the result of a misunderstanding of the intellectual purpose of some science processes and skills. With little knowledge of the scientific way of working, it was easy for a school to take the line that observation, for example, was fundamental to work in many areas of the curriculum, without necessarily acknowledging its particular significance in pattern seeking, grouping and looking for evidence in scientific enquiry. The Science 5–13 objectives included many that were essentially content-free, and this may well have encouraged some teachers, on a cursory review, to see them as contributing little to the work already taking place in their classrooms. As a result, of the already small percentage of schools that were actually acquiring this material, some were undoubtedly ignoring a structure designed to build skills and concepts progressively in favour of 'dipping in' to the topic-based materials for ideas to supplement work in progress.

Support for this view of the priorities given by teachers to work undertaken in science comes from the first of the surveys undertaken by the Assessment of Performance Unit (APU) (DES, 1981). Here, the evidence supports the view that many of the goals and emphases for science activities most warmly embraced by schools were those shared by other areas of the curriculum. In particular:

- enjoyment of science-based work;
- knowledge of the natural and physical world around;
- ability to observe carefully;
- a questioning attitude towards the surroundings.

Less stress was placed on those aspects of the scientific way of working that might be characterised as 'harder' or more specifically 'scientific', and with which many teachers clearly felt far less comfortable. These aspects included:

- ability to plan experiments;
- familiarity with correct use of simple science equipment;
- identification of variables operating in certain situations;
- recognition of patterns in observations or data;
- examining work critically for flaws in experimental method.

(DES, 1981, 37, 41)

Not surprisingly, there were found to be clear parallels between the level of performance of children in these aspects and the priority accorded them by teachers, despite the fact that there was clear evidence from the APU that children *could* plan and carry out investigations that included their appropriate application.

Moving Forward – A Continuing Focus on Skills and Processes

The early APU evidence pointed to the need for a more rigorous approach to the teaching of a scientific way of working for younger children. At the same time, the patchy nature of science teaching and learning brought many to the conclusion that there had been an over-emphasis on skills and processes fundamental to this way of working without enough consideration of the associated content. Against such a background, the 1980s brought a move towards a 'middle way' that reflected a more appropriate interrelationship between content and process. However, a sustained emphasis on the teaching of the elements of a scientific way of working remained in various initiatives to develop science in primary schools (see Harlen 1985). The consensus that emerged from the conferences that followed the Primary Survey reaffirmed the centrality of processes in any approach to develop primary science. This was to be mirrored in advice and help for schools hoping to develop a policy for science (DES, 1983), in in-service training offered by advisory services throughout the country, in changes to initial teacher training courses, in newly developed published materials for pupils such as those produced as a result of the Learning Through Science Project, and in the content of some excellent programmes for schools on radio and television. In the light of considerable pressures to give a strong focus to a transmitted content, not the least of which were political, it seems highly likely that this continuing emphasis on science skills and processes would not have been so strongly sustained without the projects, developments and perspectives of the previous twenty years.

Through the 1980s, research, curriculum development and government initiatives placing an emphasis on children's understanding and learning of skills and processes continued, and can be seen as having a direct impact on the structure of science in the first National Curriculum for Science (DES, 1989). At the same time, researchers and teachers such as Osborne and Freyberg (1985), Driver (1983) and research groups such as the Children's Learning in Science Project (CLIS, 1984–91) were calling into question a view of learning and a view of appropriate content for science that did not take into account children's own ideas and which did not give children the chance to re-structure their ideas in the light of opportunities for practical work and discussion (see Chapter 4). Some of these researchers felt, particularly with reference to science in secondary schools, that too great an emphasis had been placed on 'process science' at the expense of developing children as active learners who could modify and develop their views of science concepts through the use of the scientific way of working when appropriate. Thus, Millar and Driver strongly criticise initiatives that separated science into 'knowledge-led' or 'process-led', seeing processes as 'pedagogic means, not the ends or goals of instruction. They are … the vehicles by which children develop more effective conceptual tools and come to appreciate the types of experience where those tools may be relevant' (Millar and Driver, 1987).

Without necessarily adopting the constructivist perspective of Driver and Millar, this view is interestingly presaged by Holford (1983) in his earlier discussion of another Nuffield project that this time provided science material for the upper middle-school age range (11–13 years) during the 1970s and 1980s – the Nuffield Combined Science Project. 'The challenge is to find ways of getting a better quality of learning from the practical work … without *just* looking at a process approach to achieve this.'

As we have seen, however, the challenge was somewhat different in the primary sector. Numerous schools had seized upon newly developed work card schemes, such as Look! (Gilbert and Matthews, 1981) and Science Horizons (Hudson and Slack, 1981) as the answer to the problem of the science curriculum. For many, this did indeed raise the profile of science work within topics, though if anything resembling a constructivist approach was adopted it was in embryonic form in relatively few classes. However, from the perspective of this discussion such schemes did little, despite stated intentions, to raise the profile of science skills and processes in teaching beyond the level identified in the first APU survey findings. For this reason, publications such as *Science in Primary Schools* (DES, 1983) and *Science 5–16: A Statement of Policy* (DES, 1985) emphasised the continuing need to develop 'process science', in addition to suggesting possible areas of appropriate content for primary school science.

Influences on Practice from the Late 1980s

During the latter years of the 1980s, the question of 'content versus process' began to crystallise in response to the clear political intention to impose a National Curriculum through legislation. With a remit to work from existing 'best practice', groups such as the Task Group on Assessment and Testing (TGAT) (DES, 1988) and the Science Working Group, chaired by Professor Jeff Thompson of the University of Bath, were concerned with the overall shape and structure of the science curriculum and with its assessment. At the same time, research projects were examining practice in primary schools and beginning to suggest what might be achievable.

Significant among this research was the Science Teacher Action Research Project (STAR) which 'had the overall aim of improving practice in science education at primary school level' (Russell and Harlen, 1990). In collaboration with teachers, the research team hoped to define effective practice, to identify ways of making practice more effective and to spread such practice through in-service training. The dominant focus for this three-year project (1986–9) was an examination of the assessment of different aspects of pupils' use of scientific processes. The research team used observation, written tasks and practical tasks, rather like those devised by the APU for testing, for two main purposes. They hoped to establish the extent to which pupils understand and use process skills in science, together with appropriate and manageable methods for their assessment.

In carrying out the research, both researchers and teachers used as their focus some clearly defined processes that were, in part, based upon the Science 5–13 objectives and which were to be echoed in the structure of Attainment Target 1 (AT1) of the first National Curriculum science document (DES, 1989). For the STAR team, observing, interpreting, hypothesising, planning, measuring, recording, raising questions and critically reflecting were to be key foci of the research. Their findings with respect to assessment remain important for anyone considering appropriate means of assessing science in the classroom or through local or national testing (see Chapter 6). However, the findings that are perhaps most important from our present perspective relate to classroom practice.

The STAR team strongly supported the view that 'best practice' in primary science teaching necessarily includes an appropriate emphasis on the development of the process skills that contribute to the scientific way of working. Beyond this, however, it helped to establish some features of practice that were to receive strong emphasis in the years to come, both in the structure of the science curriculum and in emphases in initial and in-service training for teachers. Echoing the perspectives of curriculum innovations, HMI reports and DES circulars of the past twenty years, the research team made clear that the most fundamental way in which teachers might develop their practice was to become familiar themselves with the scientific way of working. In particular, it was asserted that some familiarity with how the constituent process skills could be grouped into a 'start-up skills cluster', a 'planning and doing skills cluster' and an 'interpreting cluster' would be helpful in allowing the teacher to focus teaching and assessment. Clear planning frameworks were seen to be of great value for both teaching and assessment, while the level of help that the teacher was able to offer to pupils was seen as an indicator of how successful the pupils were likely to be in their work.

With such perspectives in mind, and working hard to accommodate the political imperatives of the time, the Science Working Group for the National Curriculum produced a first report (DES, 1987) and then a series of revisions that were to culminate in the first Profile Component of the 1989 National Curriculum, titled 'Exploration of Science' (DES, 1989). It is not the intention here to repeat the account of the development of Science in the National Curriculum that is to be found elsewhere in this book, but it is clear that the dominant influence on classroom practice exerted by the National Curriculum needs to be addressed with respect to developing in pupils an understanding of the scientific way of working.

In examining the influence of the statutory requirements on practice in secondary schools, Jim Donnelly and others (Donnelly et al., 1996) provide an excellent review of the often tortuous development and sometimes tenuous hold of a 'Science 1 element' within the statutory orders. They point out numerous important features of Attainment Target 1 (AT1) in the 1989 Order that established a significant base for thinking throughout the 1990s. These features included:

- a strong influence on the structure of AT1 from the assessment framework of the APU;
- a focus on 'procedural understanding' – meaning primarily dealing with the 'number, nature and role of variables in empirical investigation' (National Curriculum Council, 1989);
- a move towards a requirement for pupils to carry out some whole investigations, though this was quite ambivalent at this stage.

While in secondary schools ensuing debates focused substantially on issues related to these features and on the place of AT17, entitled 'The Nature of Science', it is probably true to say that in primary schools the initial focus was on the fifteen ATs related to subject knowledge. Here, the spectre of lack of teacher subject knowledge caused some to doubt their ability to deliver an appropriate curriculum. With the focus given to science by its designation as a 'core' subject, in-service courses, advisory work in schools and staff meetings throughout the country focused not only on how the statutory curriculum might be accommodated within schools' long- and medium-term planning, but also on developing teacher knowledge and confidence, particularly with respect to areas of physical science. Yet in addition, these same teachers were also concerned with the requirements of AT1, not only with the need to interpret the meaning of some of the specific statements of attainment but also with what the non-statutory guidance (National Curriculum Council, 1989) called 'the nature of science itself'. In response to this the National Curriculum

Council produced training materials that were intended for use within Local Education Authorities (LEAs) to help teachers understand the distinctions between different types of practical work in science and, in particular, to emphasise the particular nature of science investigations (National Curriculum Council, 1990; National Curriculum Council, 1991). The ASE placed particular emphasis, through its publications (e.g. Summer 1989 supplement) and regional and national meetings, on helping primary teachers to come to terms with AT1.

Nevertheless, substantial difficulties lay ahead. The 1991 orders (DES, 1991) did little to clarify the meaning of the scientific way of working for teachers in primary schools, while many secondary school science teachers were becoming increasingly dissatisfied with a section of the National Curriculum orders that they regarded as unclear and for which practical implementation and assessment seemed virtually unworkable (Donnelly et al., 1996). The tortuous route to the Dearing revisions of the National Curriculum (DfE, 1995) included much confusion and substantial revision in ideas of what might constitute an appropriate science curriculum. Yet within primary schools it also undoubtedly led to a growing awareness of the nature of the constituents of scientific enquiry. The statutory nature of the National Curriculum forced all schools to consider whether their schemes of work really did address the need to develop science process skills, while the sometimes obscure nature of National Curriculum statements, together with the need to assess pupil progress in relation to them, made teachers question their own understanding of experimentation and investigation in science. In 1995 the revised National Curriculum was introduced, which included Science 1 (Sc1), 'Experimental and Investigative Science', structured in a manner remarkably similar to the way in which the STAR team recommended process skills might effectively be grouped (a 'start-up skills cluster', a 'planning and doing skills cluster' and an 'interpreting cluster'). By this time it might be argued that primary teachers had very largely equipped themselves with the understanding necessary to make sense of the components of this new structure, whether or not the reality of work in the classroom reflected the intentions of the new Sc1. Certainly, the appreciation of what constituted a fair test, the understanding of investigations as vehicles for promoting pupil decision-making and the idea of development within the learning of process skills were all ideas that were in evidence on science in-service courses, whether school or LEA based.

This is not to say that all schools and all teachers were transformed by the early experiences of working with the National Curriculum, nor that the introduction and subsequent alterations to the National Curriculum can be seen entirely as a 'force for good' on practice in schools. Indeed, it has been powerfully argued that 'the impact of the National Curriculum has been regressive, reducing the curriculum to a set of atomistic concepts to be taught' (Millar et al., 1998). Yet, with respect to the scientific way of working, it did provide a spur to developing practice that nearly thirty years of curriculum development initiatives had arguably failed to provide in many primary schools.

By the mid 1990s, what was plainly missing in many primary and secondary schools was a sense of structure to investigations that would allow pupils to gain the necessary understanding to be able to tackle novel science problems in a systematic manner. For primary schools, the work of Goldsworthy and Feasey (1994) and of some LEA advisory groups such as the Northamptonshire Inspection and Advisory Service (NIAS) (Revell, 1993) was significant in getting teachers to understand the significance of the 'planning, doing and evaluating' cycle so central to science investigations. Within the research community, significant work was being done on ideas of procedural understanding in science (it will be seen that the phrase 'procedural understanding' is not used in the same way as within the non-statutory guidance for the 1989 National Curriculum). Foremost in popularising this work have been Gott and Duggan (1995), who characterise procedural understanding as 'the thinking behind the doing', and who argue both that it is complementary to conceptual understanding and that it has to be taught. Their approach offers a rationale by which teachers might understand *why* practical work is important in science. They make it clear that developing the ability to carry out appropriate experimental work in science relies upon more than the mere acquisition of process skills or the ability to carry out relevant processes. It relies upon an understanding of the value of certain ways of collecting evidence for supporting an argument. They suggest that this understanding is comprised of 'concepts of evidence' associated with experimental design, measurement and data handling. Such concepts of evidence include, for example, understanding why it is important to repeat measurements within an experiment (see Table 5.1).

It seems likely that procedural understanding, built on a developing ability to use process skills and thus to acquire concepts of evidence, develops gradually and thus reflects the way in which conceptual understanding, built upon facts drawn from evidence, develops (Osborne and Freyberg, 1985; SPACE research reports, various dates). This important idea seems likely to have a powerful influence on the way in which teachers think of the purpose of practical work in science. Combined with structures for making investigative science work more understandable for pupils, such insights should provide a driving force for developing practical science in the classrooms of the future.

Developing Procedural Understanding

Towards the end of the 1990s, the Office for Standards in Education was noting 'Attainment in Experimental and Investigative Science (AT1) is more variable between and within schools than the other attainment targets' (Ofsted, 1997). Reviews of end of Key Stage 2 statutory assessments were concluding, with respect to pupils' ability to describe the relationship between variables, that 'just over a quarter of children on average produced full, clear answers … A further 10–15 per cent provided partial answers' (QCA, 1998a). Throughout the decade, the National Curriculum, in both its structure and its statutory assessment requirements, was seen by many as a force working against placing a proper emphasis on process skills within classroom work. We are warned to

Table 5.1: Concepts of evidence

Concepts of evidence		Definition
Associated with design	Variable identification	Understanding the idea of a variable and identifying the relevant variable to change and to measure, or assess if qualitative
	Fair test	Understanding the structure of a fair test in terms of controlling the necessary variables and its importance in relation to the validity of any resulting evidence
	Sample size	Understanding the significance of an appropriate sample size to allow, for example, for probability or biological variation
	Variable types	Understanding the difference between categoric, discrete, continuous and derived variables
Associated with measurement	Relative scale	Understanding the need to choose sensible values for quantities so that resulting measurements will be meaningful
	Range and interval	Understanding the need to select a sensible range of values of the variables within the task so that the resulting line graph consists of values which are spread sufficiently widely and reasonably spaced out so that the 'whole' pattern can be seen. A suitable number of readings is therefore subsumed in this concept
	Choice of instrument	Understanding the relationship between the choice of instrument and the required scale, range of readings required, and their interval and accuracy
	Repeatability	Understanding that the inherent variability in any physical measurement requires a consideration of the need for repeats, if necessary, to give reliable data
	Accuracy	Understanding the appropriate degree of accuracy that is required to provide reliable data which will allow a meaningful interpretation
Associated with data handling	Tables	Understanding that tables are more than ways of presenting data after they have been collected. They can be used as ways of organising the design and subsequent data collection and analysis in advance of the whole experiment
	Graph type	Understanding that there is a close link between graphical representations and the type of variable they are to represent
	Patterns	Understanding that patterns represent the behaviour of variables and that they can be seen in tables and graphs
	Multivariate data	Understanding the nature of multivariate data and how particular variables within those data can be held constant to discover the effect of one variable on another

Source: Adapted from Gott and Duggan (1995)

take care that the emphasis on process skills within the National Curriculum is not 'whittled away any more than it has been already' (Richards, 1997).

In the light of such comments it would, perhaps, be wrong to sound an over-optimistic note on the development of teaching that encourages pupils to think through how to tackle problems scientifically. Yet the progress that has been made throughout the 1990s, built as it was on developments from the 1960s onwards, can hardly be denied. International comparisons suggest that we are making acceptable progress, particularly with respect to teaching our pupils to understand fair test investigations (Harmon et al., 1997). Schools have planning in place that takes account not only of content but of the development of process skills. Science co-ordinators and specialists are working hard to ensure that there is some progression in the teaching and learning of skills in science, and there is some quite outstanding science teaching going on in every phase of schooling.

If this work is to develop there are some key issues that require attention, all of which have implications for staff training and investment, or for adaptations to the statutory curriculum and its assessment:

- Teachers need to become familiar with ideas of procedural understanding so that they can appreciate the purpose of practical work in science. Despite progress, there is still a view among teachers that practical work and skills development in science is 'a good thing' without there necessarily being a clear rationale to support this argument.
- In all future statutory orders and support materials for science produced by government sources, the centrality of Sc1 (or its equivalent) should be maintained or, better still, strengthened. This message has been at least partially acknowledged in the structure of the new statutory orders for science. But it seems likely that the structure and intentions of centrally distributed support materials for science (e.g. QCA, 1998b) will have a greater impact on practice in the classroom. With respect to such materials, the messages are less encouraging for a central focus on procedural understanding in science.
- In reviewing the early years curriculum, attention needs to be paid to any possible diminution of a scientific perspective which requires that pupils experience their world at first hand.
- Information and communications technology (ICT) is going to be of fundamental importance as a teaching and learning tool for the future. Some uses of ICT, such as data-logging, can genuinely enhance an investigative approach to science, encouraging pupils to look at the evidential value of procedures (Warwick and McFarlane, 1995). Such work needs to be given greater emphasis.
- There needs to be some stress placed on the range of possible investigations, beyond the fair test type, that can occur within the classroom. Goldsworthy et al. (1998) point out that teachers are struggling to fit all investigations within a fair test model and that this model does not necessarily fit in all cases.

If these perspectives can be taken into account, they will provide a practical basis for developing pupils' understanding of the importance of the scientific way of working.

References

Association for Science Education (1963) *Policy Statement Prepared by the Primary Schools Science Committee*, Hatfield: ASE.

Association for Science Education (1989) *National Curriculum Special Primary Science Review*, Hatfield: ASE.

Black, P. (1980) 'Why hasn't it worked?', *Times Educational Supplement*, 3 October, pp. 31–2.

Brooks, R. (1991) *Contemporary Debates in Education: An Historical Perspective*, London: Longman.

Central Advisory Council for Education [England] (1967) *Children and Their Primary Schools* (The Plowden Report), London: HMSO.

CLIS (1984–91) *Children's Learning in Science Project Reports*, Leeds: Centre for the Study of Science and Mathematics Education.

Dearing, R. (1993) *The National Curriculum and its Assessment: Final Report*, London: SCAA.

DES (1978) *Primary Education in England: A Survey by HM Inspectors of Schools*, London: HMSO, pp. 58–63.

DES (1981) *Science in Schools Age 11: Report No. 1*, London: HMSO.

DES (1983) *Science in Primary Schools*, London: HMSO.

DES (1985) *Science 5–16: A Statement of Policy*, London: HMSO.

DES (1987) *National Curriculum: Task Group on Assessment and Testing – A Report*, London: HMSO.

DES (1988) *Task Group on Assessment and Testing: A Report*, London: HMSO.

DES (1989) *Science in the National Curriculum*, London: HMSO.

DES (1991) *Science in the National Curriculum*, London: HMSO.

DfE (1995) *Science in the National Curriculum*, London: HMSO.

Donnelly, J., Buchan, A., Jenkins, E., Laws, P. and Welford G. (1996) *Investigations by Order: Policy, Curriculum and Science Teachers' Work under the Education Reform Act*, Leeds: Studies in Education.

Driver, R. (1983) *The Pupil as Scientist*, Milton Keynes: Open University Press.

Gilbert, C. and Matthews P. (1981) *Look*, London: Addison Wesley.

Goldsworthy, A. and Feasey, R. (1994) *Making Sense of Primary Science Investigations*, Hatfield: ASE.

Goldsworthy, A., Watson, R. and Wood-Robinson, V. (1998) 'Sometimes it's not fair!', *Primary Science Review*, 53: 15–17.

Gott, R. and Duggan, S. (1995) *Investigative Work in the Science Curriculum*, Buckingham: Open University Press.

Harlen, W. (1975) *Science 5–13: A Formative Evaluation*, Schools Council Publications, London: Macmillan Education.

Harlen, W. (1978) 'Does content matter in primary science?', *School Science Review*, 59: 614–25.

Harlen, W. (1985) *Teaching and Learning Primary Science*, London: Harper and Row.

Harlen, W. (1996) *The Teaching of Science in Primary Schools*, second edition, London: David Fulton.

Harlen, W., Darwin, A. and Murphy, M. (1977) *Match and Mismatch. Raising Questions: Leader's Guide*, Edinburgh: Oliver and Boyd.

Harmon, M., Smith, T.A., Martin, M.O., Kelly, D.L., Beaton, A.E., Mullis I.V.S., Gonzalez, E.J. and Orpwood, G. (1997) *Performance Assessment in IEA's Third International Mathematics and Science Study (TIMSS)*, Boston: Center for the Study of Testing, Evaluation and Educational Policy.

Holford, D. (1983) 'Nuffield Combined Science: Themes for the Eighties' in C. Richards and D. Holford (eds) *The Teaching of Primary Science: Policy and Practice*, London: Falmer Press.

Hudson, J. and Slack, D. (1981) *Science Horizons – The West Sussex Science 5–14 Scheme*, Basingstoke: Globe Education.

Millar, R. and Driver, R. (1987) 'Beyond processes', *Studies in Science Education*, 14: 32–62.

Millar, R., Osborne, J. and Nott, M. (1998) 'Science education for the future', *School Science Review*, 80 (291): 19–28.

Ministry of Education (1961) *Science in the Primary School*, London: HMSO.

National Curriculum Council (1989) *Science: Non-Statutory Guidance*, York: National Curriculum Council.

National Curriculum Council (1990) *Investigations: Working with Science AT1 in Key Stages 1 and 2*, York: National Curriculum Council.

National Curriculum Council (1991) *Science Explorations*, York: National Curriculum Council.

Ofsted (1997) *Standards in Primary Science*, London: HMSO.

Osborne, R.J. and Freyberg, P. (1985) *Learning in Science: The Implications of 'Children's Science'*, New Zealand: Heinemann Educational.

Qualifications and Curriculum Authority (1998a) *Standards at Key Stage 2: English, Mathematics and Science. Report on the 1998 National Curriculum Assessments for 11-Year-Olds*, London: QCA.

Qualifications and Curriculum Authority (1998b) *Science: A Scheme of Work for Key Stages 1 and 2*, London: QCA.

Revell, N. (ed.) (1993) *The Sc1 Book: Investigations 5–16*, Northampton: NIAS.

Richards, R. (1985) 'Learning through science' in B. Hodgson and E. Scanlon (eds) *Approaching Primary Science*, London: Harper Education.

Richards, R. (1997) 'A bumpy ride; the progress of primary science', *Primary Science Review*, 47: 3–5.

Russell, T. and Harlen, W. (1990) *Assessing Science in the Primary Classroom: Practical Tasks*, London: Paul Chapman.

Schilling, M., Hargreaves, L., Harlen, W. with Russell, T. (1990) *Assessing Science in the Primary Classroom: Written Tasks*, London: Paul Chapman.

SPACE (various dates) research reports, e.g. T. Russell and D. Watt (1990) *SPACE Research Reports: Growth*, Liverpool: Liverpool University Press.

Steadman, S. (1978) *An Enquiry into the Impact and Take-up of Schools Council Activities*, Schools Council Publications, London: Macmillan Education.

Warwick, P. and McFarlane, A. (1995) 'IT in primary investigations', *Primary Science Review*, 36: 22–5.

Wastnedge, R. (1983) 'Nuffield Junior Science: The end of a beginning?' in C. Richards and D. Holford (eds) *The Teaching of Primary Science: Policy and Practice*, London: Falmer Press.

6 Assessing Learning in Science

Rachel Sparks Linfield and Paul Warwick

Over the years much has been written and discussed about assessment. The introduction of the National Curriculum in science emphasised the need for teachers to assess children against predetermined criteria. Teachers rapidly had to become proficient in using as many modes of assessment as possible, in order to collect evidence for making end of Key Stage judgements. Many teachers became swamped by the paperwork, while the outside pressures from Ofsted inspections and league tables outlining test results created the impression that assessment was primarily for purposes of external scrutiny. Thus, the most important reason for carrying out assessments, namely to enhance and monitor children's learning and understanding, took second place as concerns for teacher accountability to external agencies took over.

This chapter will explore the way assessment of science has developed and will focus on the modes of assessment most frequently used by teachers. It will examine the impact on assessment by bodies such as the Assessment for Performance Unit (APU) and suggestions will be made as to how assessments in science might meaningfully be carried out with future generations of children.

What is Assessment?

Harlen (1996) defines assessment as

> a word used both for the *process* of making judgements about pupils' achievement and for the *product* of this process, that is, the judgement which is made. Judgement is always involved in assessment, since it is a process in which the actual behaviour is replaced by some description, sign or number.
>
> (Harlen, 1996, p. 148)

Thus, it can be seen that when carrying out assessments in science, teachers must consider both how to assess and what to assess. In addition, they must give thought to the reasons behind carrying out the assessment, and here it is useful to think about two purposes identified within the National Curriculum Task Group for Assessment and Testing Report (DES, 1987), namely:

formative, so that the positive achievements of a pupil may be recognised and discussed and the appropriate next steps may be planned;
summative, for the recording of the overall achievement of a pupil in a systematic way.

(DES, 1987, para 23)

Teachers have always carried out formative assessments, albeit in an informal way. Whenever, for example, a child is observed measuring, recording or talking, judgements are being made as to the standard at which the child is performing and where the child needs to be directed next. It is, however, the planning for such observations that helps a teacher to make quality assessments. While occasionally it is good just to stand back and simply observe without an agenda, at other times particular skills and concepts will need to be systematically observed and recorded. At such times it is important to consider, in relation to criteria, what constitutes evidence of achievement and therefore what a child might be expected to demonstrate, say or produce as some form of permanent record. Only then can assessments be made in an objective, measurable way and 'appropriate' targets be planned for.

Summative assessments in science, particularly at the primary level, are a more recent practice heavily influenced by the National Curriculum end of Key Stage tests. This is demonstrated within the extracts taken from four 10-year-old children's annual school reports over four decades (see Figure 6.1). The earlier reports reflect subjective comments, suggestive of little, if any, formal summative assessment.

It is interesting to note that in the 1960s, where science was taught at all in primary schools, it was usually as part of a topic. In the first two extracts, teachers merely comment on pupils' interest and content areas. In the latter two, skills are spoken of and comments go beyond the enjoyment factor. It is in the final comment that, due to National Curriculum requirements, the summative judgement about Emma's ability in science includes a quantitative element.

The reports highlight the fact that in order to make judgements about a child's progress in science it is necessary to consider both conceptual and procedural understanding. Science is often broken down into concepts and skills. It is, however, important when assessing children's capabilities in science to consider not only whether they can carry out a skill or process but, in addition, whether they understand why they need to use it. Emma's report clearly reflects this procedural understanding. This final report also emphasises the time-consuming nature of assessment. In order to assess, teachers need to use a range of methods which will encompass knowledge, skills and, most importantly, understanding. They need time to plan and to reflect. As Ritchie (1998) states, 'Effective assessment requires effective classroom management. To assess, teachers need to spend "quality" time with individuals and groups. This is only possible in a well-organised and managed classroom.' Part of this 'effective management' must clearly be wide knowledge of the various modes of assessment a teacher might employ. The following examples are not exhaustive, but consideration has been given to the forms most commonly and successfully

Figure 6.1: Extracts from annual reports

July 1964
Topic: Elizabeth shows interest. She enjoyed collecting wild flowers and brought in many for our nature table.

July 1971
Science and environmental studies: Lucy enjoys experimenting. She particularly liked making a circuit to light a bulb and working with magnets.

July 1988
Science: This term Jonathan has investigated materials, forces and plant growth. He shows the ability to observe carefully in a critical way and to measure accurately. He understands how to design and carry out fair tests.

July 1997
Science: Emma has achieved a high standard in all the areas of science covered this year. She particularly enjoyed working with plants and investigating sound and music. She capably plans and carries out experiments and understands the need to repeat measurements and to carry out fair tests.
National Curriculum Teacher Assessment: Level 5
National Curriculum Test: Level 5

used in schools in the assessment of both conceptual and procedural aspects of children's learning in science.

Modes of Assessment

Observation of Children at Work

Clayden and Peacock (1994) stress the importance of observation 'with a purpose' in the assessment of both conceptual and procedural understanding in science. It will be apparent that although written reports produced by some children may give evidence of, for example, the ability to plan and record, many children are unable to give a true picture of their understanding. This is particularly true in the assessment of early years science.

The Science Teacher Action Research (STAR) Project carried out by Liverpool and Leicester Universities during the late 1980s pointed to the vital role of classroom observation in the assessment of pupils' process skills in science. Believing that 'no one method of assessment enjoys sufficient advantages over others to ensure its exclusive use' (Cavendish et al., 1990), the research team looked at ways of encouraging teachers to take a more structured approach to observing in the classroom. Observation schedules were created for research purposes, and were later modified by teachers to allow a more criterion-referenced observation of pupils' use of process skills to be undertaken in their day-to-day teaching

Several issues emerged from this research. Perhaps the most important was that classroom observation for assessment is always 'a compromise between what is practical and what is desirable' (Cavendish et al., 1990). The research showed that the quality of observation was strongly affected by the quality of teachers' planning. Where learning objectives were tightly defined then the teacher was much more likely to be able to make accurate judgements about a child's progress. The need for teachers to have evidence to support assertions was strongly emphasised. This focus on the need for evidence encouraged teachers to take responsibility where pupils experienced difficulties rather than attribute failure to factors outside the control of the teacher, such as a pupil's home background.

Significantly, teachers' understanding of the scientific way of working was found to be central to the making of informed assessments of their pupils' development of process skills. This is a theme returned to throughout this book. Here, it is important to note that, as a precursor to the introduction of the National Curriculum, the STAR team felt it was important for teachers to have an understanding of the planning, carrying out and interpreting cycle involved in science investigations.

Classroom observation can clearly provide insights into pupils' use of appropriate concepts within the structure of practical work, but it is primarily through other modes that the teacher gains an insight into a child's conceptual understanding in science.

Assessing the Products of Children's Work

The Primary Science Process and Concept Exploration (SPACE) Reports (e.g. Watt and Russell, 1990) illustrate how useful drawings can be for assessing children's understanding of concepts. It is clear, however, that teachers should be wary of assuming too much when regarding products alone. All too often a drawing which appears to reflect understanding to the adult eye is actually not what the child intended. Equally, a drawing without detail may reflect a child who has gone beyond the understanding anticipated by the teacher. Take, for example, Edmund, aged 6, who when asked to observe and draw a straw in water simply drew a straight straw. The teacher thought that Edmund had failed to see that the straw in water appeared to bend. In fact, Edmund had noticed this but also had realised that the straw was in fact still straight. He stated when asked why he had not drawn it bent, "Cos it's only pretend. Really the straw is straight" (Sparks Linfield and Warwick, 1996). Thus, in this case the product gave misleading evidence of the child's understanding. Discussion was necessary to gain a full picture of Edmund's knowledge of the optical effect.

Nevertheless, the assessment of products of children's scientific endeavours is widely favoured by teachers. It provides evidence which can be kept and allows teachers time to form judgements. The SPACE Reports clearly indicate that drawings can form a powerful bridge, helping the teacher to access children's understanding and providing an indication of development. An example of the use of drawings to indicate very different development in children of the same age is illustrated in the pictures, answering the question, 'What is inside your body?' (See Figure 6.2.)

Concept maps as a means of assessing children's knowledge are now fairly well established within the repertoire of teachers' assessment of science products. Novak and Gowin (1984) established the idea of concept mapping as a means of discovering the 'propositional links' that children make between concepts. Even very young children are capable of producing these 'maps' and showing what they understand about the relationship between concepts. (See Figure 6.3.)

The maps in Figure 6.3 are produced by two children in a reception class who had carried out practical experiments on making ice and floating freezer-made 'icebergs'. The analysis of these maps suggests that Fraser has a beginning concept of the interrelationship of states of matter, whereas Sam simply describes what he knows about water and ice. This understanding was confirmed by the class teacher in discussion, illustrating the intimate relationship between modes of assessment.

Perhaps the greatest benefit of this technique is for those who teach holding a constructivist view of learning (see Chapter 4, and Ollerenshaw and Ritchie, 1993). If it is to be successful as a tool to analyse prior knowledge or to evaluate learning resulting from a unit of work, it will be clear that:

> children need to be taught this technique and need to understand particularly how important the 'joining words' are in making the whole thing have meaning. As with all methods used for recording children's thoughts it should not be used too frequently. This is not just to prevent fatigue on the part of the children; concepts maps are time-consuming for the teacher to analyse and should be seen as only one tool in the teacher's assessment armoury.
>
> (Sparks Linfield and Warwick, 1996, p. 93)

Discussion and Questioning

There is a wealth of research literature which relies heavily upon interviewing, more or less formally and often linked to the carrying out of practical tasks, to

Figure 6.2: Responses to the question, 'What's inside your body?'

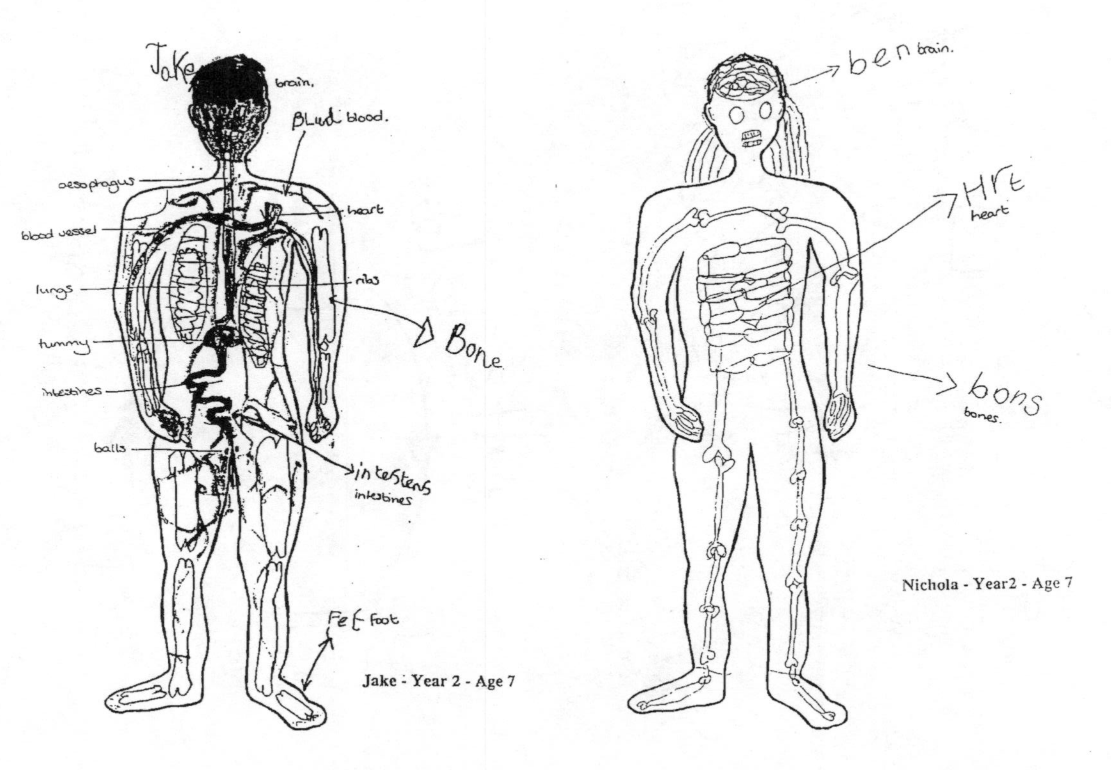

Jake - Year 2 - Age 7

Nichola - Year2 - Age 7

Figure 6.3: Concept maps by Sam and Fraser

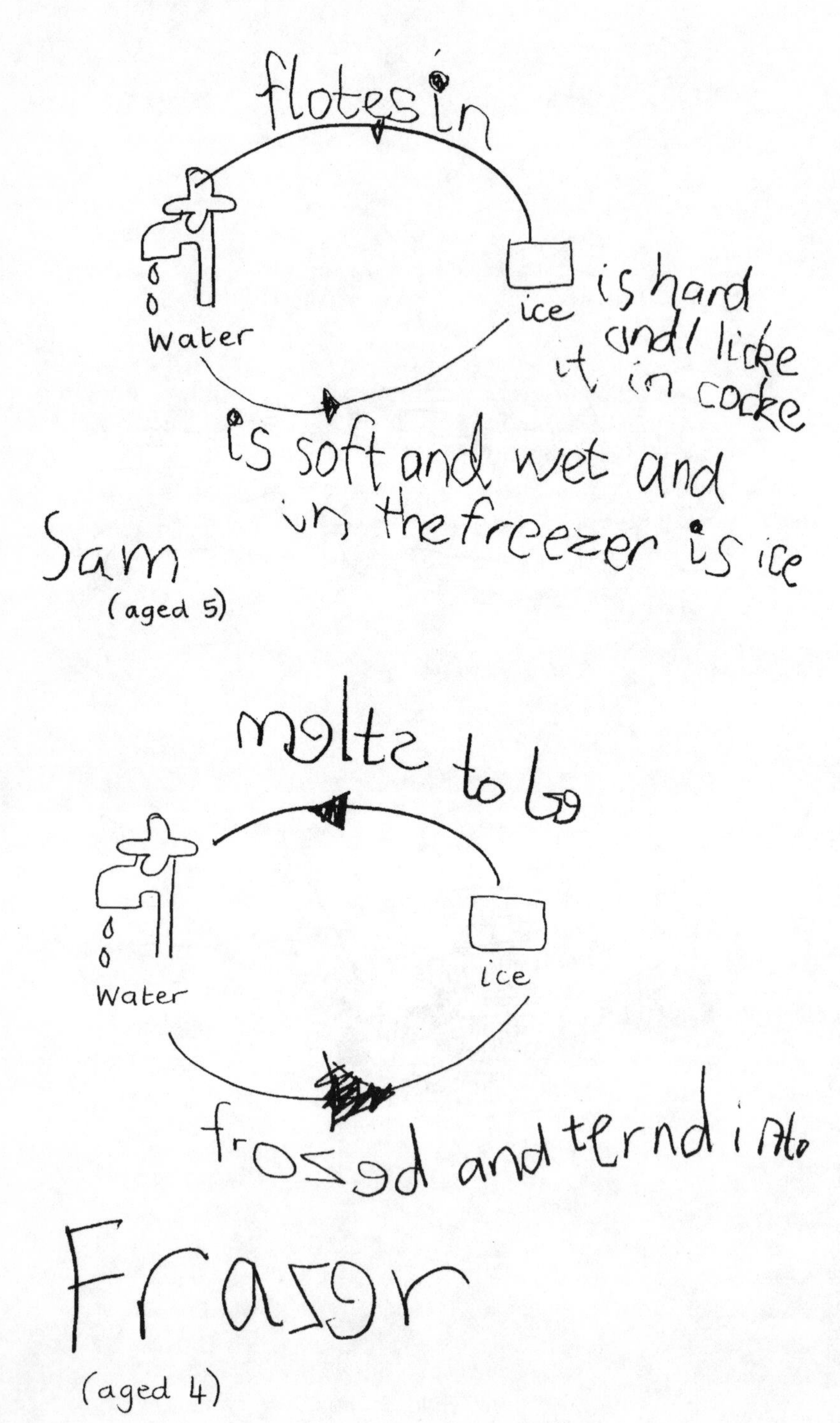

ascertain children's ideas in relation to science concepts (Inhelder and Piaget, 1958; Osborne and Freyberg, 1985; Russell et al., 1998). This reflects strongly the most powerful assessment mode in the teacher's armoury – the ability to investigate understanding through discussion and questioning. Feasey and Thompson (1992) show how asking questions which sometimes focus either on practical activities or on products can enable children to move forward in their understanding as well as providing a valuable aid to teacher assessment. In addition, discussion can enable children to become involved in the assessment of their own work and thus help them to become autonomous learners.

Most teachers will recognise that the majority of pupils are better able to express both conceptual and procedural understanding through talk rather than through writing. This may primarily be the result of the facility that pupils have for using the two modes of expression. In addition, both pupils and teachers may have different agendas for what it is appropriate to address in speech and in writing. Research by Warwick et al. (1999) graphically illustrates that, with respect to procedural understanding, the written work linked to practical tasks in science rarely reflects children's true understanding. Compare, for example, the written work produced by Alice (aged 11) with her discussion of the same experiment (see Figures 6.4 and 6.5).

The teacher rarely has the luxury of spending this amount of time with a single pupil, but this example clearly illustrates the value of discussion and questioning for probing children's scientific understanding.

External Influences on Assessment Practice in School

Since the introduction of Science in the National Curriculum in 1989 (HMSO, 1989) teachers have worked hard to incorporate the various modes of assessment into their classroom practice. The importance of having evidence to support criterion-based judgements has been paramount. It is useful at this point, therefore, to consider the influences which have informed current assessment practice within the framework of the National Curriculum.

The work of the Assessment of Performance Unit is also considered in Chapters 2 and 5. Here, however, it is important to emphasise the influence which the APU exerted in demonstrating how schools could assess science processes and investigations. Their remit was wider than the assessment of factual knowledge or children's understanding of science concepts. The APU carried out national surveys in England, Northern Ireland and Wales between 1980 and 1984. Surveys used an extensive bank of practical and written test items which covered concepts, skills and investigations. The APU recognised that assessment is context-dependent and took care to assess skills through a range of conceptual areas. It also acknowledged that while it was important to find out whether children could use a variety of skills, it was equally vital to ascertain if the same skills could be used proficiently within the context of whole investigations.

The results of the surveys were written up as a series of reports (e.g. DES, 1983). The reports outlined how children in the varying age groups had

Figure 6.4: Alice's written work

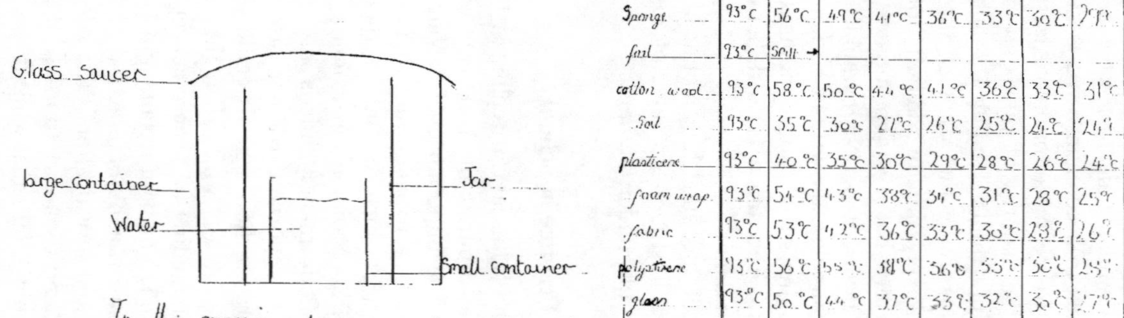

Glass saucer

large container

Water

Jar

Small container

In this experiment we are trying to find out whats the best insulator to keep the water warm. We are putting different materials around the pot. Then we are going to put some boiling water in the pots. The temperature of the water will be about 80 degrees. There are nine pots and the materials in the pots are foil, cotton wool, sponge, soil, plasticine, foam wrap, fabric, polystyrene and glass. We are also going to test our water every 15 minutes.

We have found out that cotton wool is the best insulator and you can look at the results on the chart. ✓ very good

The Results of our insulation Experiment.

Materials	Times → 1.35	1.50	2.05	2.20	2.35	2.50	3.05	3.20
Sponge	93°C	56°C	49°C	41°C	36°C	33°C	30°C	[illegible]
foil	93°C	Spilt →						
cotton wool	93°C	58°C	50°C	44°C	41°C	36°C	33°C	31°C
Soil	93°C	35°C	30°C	27°C	26°C	25°C	24°C	24°C
plasticene	93°C	40°C	35°C	30°C	29°C	28°C	26°C	24°C
foam wrap	93°C	54°C	43°C	38°C	34°C	31°C	28°C	25°C
fabric	93°C	53°C	42°C	36°C	33°C	30°C	28°C	26°C
polystirene	93°C	56°C	55°C	38°C	36°C	35°C	30°C	25°C
glass	93°C	50°C	44°C	37°C	33°C	32°C	30°C	27°C

Figure 6.5: Alice's discussion of her work

We had to do some things about insulation and it's to keep the water warm and how we did it was we got a jug and we put a little pot of water in it, and a plate over the top, and we put it in a large container and we tested it and we were seeing how long it would take to cool down. We were seeing how different materials would affect how warm it would stay.

What did you change?
We changed the different materials that we put in to it, so it could be cotton wool or something.

What did you keep the same and why?
We kept the same pot, the same amount of water. It was important because otherwise it wouldn't be a fair test.

How many materials did you use and why did you use this number of materials?
Nine different materials were each tested once, to find out which was the best insulator.

What units did you use for your measurements?
Degrees Centigrade.

Why did you draw your graph like this?
We discussed it and decided to put the times down here [horizontal axis] and the temperature up here [vertical axis].

What did you measure and how did you do it?
We measured the time, the temperature of the water, every 15 minutes. We had to do each one at the same time or it would have had longer to cool down. We used degrees Centigrade.

Did you repeat the test?
The test wasn't repeated. The results might have been different if the temperature in the room was different.

How accurate were your results?
They might have been more accurate if the test was repeated.

If you repeated the test, how would you know which results to use for the graph?
You'd use the ones that came up most times.

continued

Why did you use a results table designed before the experiment?
So that the results could be recorded as they were observed. The graph was designed after, once we knew what we had to show.

What type of graph did you draw? Could you have used a different type of graph?
We made a line graph of the most successful ones that stayed the hottest. The results of all nine materials were also recorded on a chart, given to us by the teacher. We could have done a bar chart but a line graph is clearer so that you can see what's going on.

What do the graph and table tell you about your results?
You can see how much time it took down here and where the hottest was up here.

performed and provided valuable information for teachers, both in terms of implications for teaching and in ideas for how to carry out assessments. It is the latter which possibly was of most benefit to practising teachers. Hayes writes:

> Investigative skills are complex and the degree of difficulty increases significantly if these are assessed through concepts at a corresponding level of complexity. It is often more effective to assess these through less complex concepts in order that the focus of assessment is directed at the investigative work.
>
> (Hayes, 1998, p. 126)

In providing a range of topic areas and contexts the APU did this effectively. Indeed, an investigation to compare balls bouncing on a variety of surfaces and used within the National Curriculum as an optional task at Key Stage 1 (KS1) (SCAA, 1994) bears remarkable similarity to the one described within the APU's *Science at Age 11* report (APU, 1988). When the National Curriculum was first introduced, KS1 teachers were extremely vocal over the need for assessments to be meaningful, to be practical and for them to be able to be carried out as part of the normal classroom work. The first Standard Assessment Tasks (SATs) for KS1 in 1991 did take account of these desires. Teachers were able to choose tasks from a range of content areas, so that in theory the assessments would fit into their class topics. For example, a Year 2 teacher whose class was studying 'Water' might have chosen to assess her class on 'Floating and sinking', while other classes would sort materials according to their properties.

Yet the idea of producing structured investigations for assessment tasks is by no means the only way in which the work of the APU has influenced science assessment. The research demonstrated that it was possible to assess some process skills through written tasks. In particular, written tests were used for the assessment of reading information from tables, graphs and charts, interpreting presented information and planning investigations (Russell et al., 1988). This was reflected in the later work of the STAR research team, which again emphasised the importance of a meaningful context for testing children's capabilities in science. Here, written test items and practical tasks were created to assess process skills within the context of investigations linked to specific concept areas (Schilling et al., 1990; Russell and Harlen, 1990). It would appear that such work influenced the report of the Task Group on Assessment and Testing (DES, 1987) and fed into the structure of statutory assessment at the end of Key Stages 1, 2 and 3. The TGAT authors stated strongly that assessment 'should be an integral part of the educational process, continually providing both "feedback" and "feedforward". It therefore needs to be incorporated systematically into teaching strategies and practices at all levels' (DES, 1987, para. 4). In addition, however, they were clear that 'Performance would be assessed using a combination of informal assessment ... and the formal standardised tasks set on a national scale' (DES 1987, Appendix L, para. 9).

It is these twin pulls, the need to hone classroom assessment procedures and the requirement for formal assessment at the end of the Key Stages, that drove assessment practice throughout the 1990s.

The first SATs for science at Key Stage 1 were very much influenced by the desire to make statutory assessment as much a part of everyday classroom practice as possible. They involved practical situations, but it was quickly found that the actual tasks involved a lot of one-to-one work, during which other children had to be given more mundane activities which, in theory, could be completed independently. Thus, when the first science SATs were evaluated concern was shown for the amount of time and resourcing they entailed. As a result, successive years' end of Key Stage statutory assessments involved less and less practical work, until finally they were no longer a requirement at KS1. (Only English and Mathematics are now tested at KS1.) With the introduction of statutory tests at the end of Key Stage 2 in 1995, after two years of pilots, there was no attempt made to include practical science tasks or investigations.

It might be argued that to exclude practical tasks from statutory assessment was a retrograde step. Undoubtedly the decision to do this was more than partly a pragmatic one, yet it could be argued that the written tests are not blunt instruments. In some ways they mirror those of the APU and STAR, in that 'the attempt is made to set questions in contexts which represent the best practice in KS2 Science classrooms; exploratory and investigative contexts are introduced using real data which are pertinent to the question, for example' (Schilling, 1998, p. 133).

This contrasts somewhat with teachers' initial experiences with statutory testing. With the introduction of Science in the National Curriculum, it would probably be true to say that teachers felt divorced from an assessment process imposed upon them. With generally low levels of confidence in science, many primary teachers felt threatened by the need to collect evidence of children's attainment. Recording, through the ticking of seemingly endless boxes corresponding to the acquisition of ideas and skills, tended to obscure the idea that this information was intended to inform the next stages of teaching and learning. Further, the introduction of end of Key Stage tasks and tests sometimes had the effect of undermining confidence in the assessment process as a whole. Clearly, in order for quality judgements to be made on the basis of National Curriculum tests, the test items themselves needed to be valid and the markers had to be aware of 'child speak'. In the early KS1 SATs, one item asked children to draw themselves using a force to change the shape of a can and to describe what they were doing. One child drew a 'v' behind the can and wrote, 'I am sticking up two fingers to change the can into a bunny.' Sadly the question did not take account of the audience. To many children, 'changing the shape' is a phrase used in drama to mean 'pretend to be a ...' Thus, the question gave the wrong cues. Furthermore, the marking script did not allow the child to receive a mark, despite the fact that to hold two fingers up does require a force!

In recent years it does seem that teachers have developed their expertise in using a range of modes of assessment which enable them to make increasingly informative judgements about the scientific capabilities of their pupils. What seems to have happened during the 1990s is a combination of three key factors:

- most teachers have become rather more confident in their understanding of the requirements of the National Curriculum, partly helped by the moderation of children's work carried out within schools and by the use of published exemplification materials from the School Examinations and Assessment Council (SEAC, 1993) and the School Curriculum and Assessment Authority (SCAA, 1995);
- many teachers have increased their understanding of how the modes of assessment might be used to set targets for future learning in science;
- the 'Dearing Review' (Dearing, 1994) has made explicit the fact that teachers are not required to keep a mass of evidence of attainment.

The government's response to the Dearing Review of 1993 made it clear that a revision of the National Curriculum could create 'more opportunity for teachers to use their professional skill; it would be left up to teachers to decide how to make and record assessments of pupils' classroom work in ... science' (Dearing, 1994). Further, in his review Dearing notes evidence from the Office for Standards in Education (Ofsted) that 'Teacher assessment and statutory testing have both played a part in improving teachers' understanding of the National Curriculum and of the standards that are expected' (ibid.). This is borne out by the fact that most teachers now have a 'notion of levelness' in

Table 6.1: Standards in primary science

	Percentage of schools at Key Stage 1 graded satisfactory or above	*Percentage of schools at Key Stage 2 graded satisfactory or above*
Quality and use of day-to-day assessment	70%	65%
Procedures for assessing pupils' attainment	75%	70%
Use of assessment to inform curriculum planning	55%	50%

Source: Derived from Ofsted (1997)

science, helping them to understand the broad attainment of the children they teach.

Ofsted have reported, in their review for the academic year 1996–7, that teachers' informal day-to-day assessment usually provides a clear and accurate review of pupils' relative achievement' (Ofsted, 1997). The report indicates the percentage of teachers who are satisfactory or above with respect to the assessment of primary science (see Table 6.1).

While these findings give no cause for complacency, they do suggest that the majority of teachers are effective information gatherers who have come a long way since the inception of the National Curriculum, but that more work needs to be done in helping teachers to use assessments to inform planning.

Looking to the Future

When discussing the skills necessary for teachers to have to assess science, Ritchie indicates that these involve

> focused observation, active listening, responsive questioning and the skill of recognising significance amongst the overwhelming amount of information that a teacher can potentially gather in a classroom concerning children's learning. The quality of the judgements that teachers make is the key to effective assessment.
>
> (Ritchie, 1998, p. 117)

It is interesting to note that his list includes many action words involving teachers and children working together. Drummond supports this view and goes further by insisting that 'The process of assessing children's learning – by looking closely at it and striving to understand it – is the only certain safeguard against children's failure, the only certain guarantee of children's progress and development' (Drummond, 1993, p. 10).

As we noted at the end of the last section, it is this latter aspect, the use of assessment to enhance children's learning, that is so vital and yet comparatively neglected. Teachers have clearly moved from the panic 'tick box' response to assessment demands to a more considered view of what and how evidence is to be collected and collated. It seems that perhaps what is needed to move assessment in science forward is a more diagnostic focus 'through which learning difficulties may be scrutinised and classified so that appropriate remedial help and guidance can be provided' (DES, 1987). The TGAT Report recommended 'that a review is made of the materials available for detailed diagnostic investigation of pupils' learning problems' (DES, 1987). Unlike in English and mathematics, however, there is little evidence of diagnostic assessment and the provision of additional help for specific children in science. Why might this be the case?

There is still a question of teacher confidence in science. Chapters 4, 5 and 7 indicate that some teachers need to develop their own conceptual and procedural understanding in order that they may help their pupils to move forward. Furthermore, information about children's common misconceptions in science, which would allow teachers to make informed judgements about progress, have only recently been given prominence within initial teacher training courses (ITT). Probably most importantly, commercial diagnostic materials for such testing are not readily available. End of Key Stage testing does provide broad perspectives against which individual performance can be judged in the future. For example, in the analysis of children's performance in the 1998 National Curriculum assessments for 11-year-olds provided by the Qualifications and Curriculum Authority (QCA), we are told that 'children should practise the use of conventional forms in representing scientific knowledge', that 'the major stages in the life cycles of flowering plants appear not to be widely understood' and that 'confusions continue to surround children's appreciation of the magnetic and conducting properties of metals' (QCA, 1998a). Such statements can be helpful to teachers but only partially contribute to their ability to make diagnostic assessments of individuals in science. Similarly, the QCA scheme of work for science in KS1 and 2 (QCA, 1998b) provides a range of more specific benchmarks, in terms of learning objectives and anticipated learning outcomes, against which individual performance can be compared. This is still, however, a 'broad brush', with teachers being encouraged primarily to identify those children whose 'progress differs markedly from that of the rest of the class' (QCA, 1998b). No advice is given about what to do where this is the case. Although this is somewhat disappointing, it is not surprising in a curriculum that is placing increasing emphasis on numeracy and literacy. The greater emphasis on individual needs within the new statutory orders for science should, however, promote the development of diagnostic assessment within science for all children.

Perhaps the future rests on the extent to which British classrooms fully embrace the technological revolution, exemplified by the vision of the potential of the National Grid for Learning and examined in some detail in Chapter 8. To make meaningful assessments in science, teachers require not only an in-depth knowledge of both assessment and science, but also quality time to plan and to spend with pupils, quality resources and the kind of quality assistance that a fully on-line classroom might provide through help with planning, diagnosis and teaching.

References

APU (1988) *Science at Age 11. A Review of APU Survey Findings 1980–84*, London: HMSO.

Cavendish, S., Galton, M., Hargreaves, L. and Harlen, W. (1990) *Assessing Science in the Primary Classroom: Observing Activities*, London: Paul Chapman.

Clayden, E. and Peacock, A. (1994) *Science for Curriculum Leaders*, London: Routledge.

Dearing R. (1994) *The National Curriculum and its Assessment: Final Report*, London: SCAA.

DES (1983) *Science Report for Teachers 1: Science at Age 11*, London: HMSO.

DES (1987) *National Curriculum: Task Group on Assessment and Testing – a Report*, London: HMSO.

Drummond, M. (1993) *Assessing Children's Learning*, London: David Fulton.

Feasey, R. and Thompson, L. (1992) *Effective Questioning in Science*, Durham: Durham School of Education.

Harlen, W. (1983) *APU Science Report for Teachers: 1, Science at Age 11*, London: Assessment of Performance Unit.

Harlen, W. (1996) *The Teaching of Science in Primary Schools*, London: David Fulton.

Hayes, P. (1998) 'Assessment in the classroom', in R. Sherrington (ed.) *ASE Guide to Primary Science Education*, Hatfield: Association for Science Education, pp. 125–9.

Inhelder, B. and Piaget, J. (1958) *The Growth of Logical Thinking from Childhood to Adolescence*, London: Routledge and Kegan Paul.

Novak, J. and Gowin, D. (1984) *Learning How to Learn*, Cambridge: Cambridge University Press.

Ofsted (1997) *Standards in Primary Science*, London: HMSO.

Ollerenshaw, C. and Ritchie, R. (1993) *Primary Science: Making It Work*, London: David Fulton.

Osborne, R. and Freyberg, P. (1985) *Learning in Science: The Implications of Children's Science*, London: Heinemann.

QCA (1998a) *Standards at Key Stage 2. English, Mathematics and Science. Report on the 1998 National Curriculum Assessments for 11-Year-Olds*, London: QCA.

QCA (1998b) *A scheme of work for Key Stages 1 and 2: Science*, London: QCA.

Ritchie, R. (1998) 'Implementing assessment and recording as a constructive process', in A. Cross and G. Peet (eds) *Teaching Science in the Primary School Book Two: Action Planning for Effective Science Teaching*, Plymouth: Northcote House.

Russell, T. and Harlen, W. (1990) *Assessing Science in the Primary Classroom: Practical Tasks*, London: Paul Chapman.

Russell, T., McGuigan, L. and Hughes, A. (1998) *SPACE Research Reports: Forces*, Liverpool: Liverpool University Press.

Russell, T., Black, P., Harlen, W., Johnson, S. and Palacio, D. (1988) *Science at Age 11: A Review of APU Survey Findings 1980–84*, London: HMSO.

SCAA (1994) *Scientific Investigation: Optional Task Materials to Support Teacher Assessment in Sc1 at Key Stage 1*, London: HMSO.

SCAA (1995) *Consistency in Teacher Assessment. Exemplification of Standards. Science: Key Stages 1 and 2, Levels 1 to 5*, London: SCAA.

Schilling, M. (1998) 'KS2 Standard Tests in Science', in R. Sherrington (ed.) *ASE Guide to Primary Science Education*, Hatfield: The Association for Science Education, pp. 130–3.

Schilling, M., Hargreaves, L., Harlen, W. with Russell, T. (1990) *Assessing Science in the Primary Classroom: Written Tasks*, London: Paul Chapman.

SEAC (1993) *Children's Work Assessed: English, Mathematics and Science*, London: SEAC.

Sparks Linfield, R. and Warwick, P. (1996) ' "Do you know what MY name is?" Assessment in the early years: some examples from science', in D. Whitebread (ed.) *Teaching and Learning in the Early Years*, London: Routledge, pp. 81–98.

Warwick, P., Sparks Linfield, R. and Stephenson, P. (1999) 'A comparison of primary pupils' ability to express procedural understanding in science through speech and writing', *International Journal of Science Education*, 21 (8): 823–38.

Watt, D. and Russell, T. (1990) *Sound: Primary SPACE Project Research Reports*, Liverpool: Liverpool University Press.

7 Views on the 'Right Kind of Teacher' for Science 3–13

John Hobden

Can there be a 'right' kind of teacher for science? There is not always a right kind of answer to many of the scientific questions posed for children to investigate and so it may be that there is no single 'right' kind of teacher for the subject. In the last thirty years we have witnessed rapid changes to the way science and education is viewed, and it would be surprising if views on the right kind of teacher for science had not changed too. This chapter will seek to examine some of the characteristics of good teachers of science, track changes in approach to the training of science teachers and illustrate how science teaching might be taken forward by bringing science teachers and their pupils closer to the real world of scientists.

So, how important is the teacher in bringing science to children? Ponchaud (1998) considers that 'teaching is the most significant of the factors which contribute to high standards in science education'. This conclusion derives its authority from a considerable amount of data gathered during school inspections, where the quality of teaching and the level of pupil attainment have been judged in a climate generally perceived to be hostile to teachers. Good teaching and the characteristics of good teachers therefore deserve careful analysis.

What is 'right' for providing the essential experiences of early years education may be very different from the model of the teacher of Key Stage 2 unravelling children's alternative frameworks, or the Key Stage 3 teacher introducing those parts of scientific experience which are only capable of understanding by children with formal operational thought. The right kind of teacher for the small rural primary school may not be the same as for a large one or for Key Stage 3 of a secondary school. Moreover, some teachers may be charged with the crucial tasks of co-ordinating activities over more than one Key Stage. It is these teachers who will need the greatest reserves of personal resources, both in subject knowledge and in pedagogical terms. They will need to be flexible, adaptable and sympathetic to the concerns of colleagues. They may be enthusiastic about science but will need to channel that enthusiasm in ways which will support less confident or beginning teachers. They must be both authoritative and deserving of emulation. A more general guiding principle, drawing an analogy with explanations of the nature of science itself suggests that the right kind of teacher is the one who is most able to take the child from the level at which he understands the world to something closer to the latest understanding of the adult scientific community.

Four Decades of Change

The move in primary science from what was essentially nature study to an approach which attempts to give pupils greater understanding through structured active experiences has not been without tensions. Many of the same conflicts relating to child-centred versus subject-centred education have been played out in every subject of the curriculum, but with one major difference in slant in the case of science. While the so-called 'trendy' teachers of the 1960s and 1970s were developing 'real' science of a sort, often piecemeal and without structure, there were at the same time initiatives such as the West Sussex Science Horizons Scheme (Hudson and Slack, 1981) making some limited headway towards developing pupils' scientific process skills by encouraging children to gain an insight into science as a way of working. The effective use of this approach was hampered somewhat by insufficient understanding on the part of the teachers of the time, a situation which would not be fully remedied for many years. The primary school 'nature table' of the 1950s and before provided a useful ad-hoc drip feed of biological experience, with observation being encouraged even if explanation was given as fact instead of allowing hypothesis and further structured investigation. Secondary school science was, by contrast, highly structured but still geared to the transmission of facts backed up by 'experiments' designed to 'prove' the accepted wisdom.

During the 1970s, with the encouragement of the Association for Science Education, a number of schemes were proposed which were to lay the foundations for investigative science in the primary classroom. An example of such a pioneering scheme is the Sussex *Science Horizons* (1981) material (re-published in the 1990s as *New Horizons*) which set out a range of practical investigations through which children could come to an understanding of the accepted scientific explanation for a certain phenomenon. The scheme was supported by a range of specially designed equipment designed to facilitate these investigations and thus relieve the non-specialist teacher of much of the burden of preparation. In the hands of skilled science teachers this kind of scheme led to the development of useful and appropriate skills which served as a good preparation for the work of the secondary school. However, the secondary school was not, and in some cases is still not, prepared for an influx of pupils at Year 7 possessing some ability to design their own investigations. Indeed, this clash of culture has become, if anything, more noticeable since the introduction of the National Curriculum made the acquisition of questioning and investigative skills part of the mandatory requirements (DES, 1989). Numerous calls have been made for greater continuity of approach at this particular stage of transition (e.g. Jackson, 1998, Goldsmith, 1998). Ponchaud (1998) finds that liaison between primary and secondary schools has improved but 'All too often the achievements of pupils are not fully recognised or built upon early in their secondary experience.' This finding should not come as a surprise since secondary schools need a common place to start, with an intake comprising pupils with a very varied background in science. Jarman (1993) found that many pupils, after transition to secondary school, considered that primary science was not 'real' science. She argues for a consistency of environment and teaching approach to ensure that primary and secondary school science are valued equally, a situation which will not come about under the current funding formula. The writer knows of one example, however, where a primary school teacher worked one day each week teaching Year 7 in the secondary school fed by her primary. This would seem to offer a way to solve the approach side of things but does not address the funding issue.

The inclusion of science as a core subject in the primary school curriculum meant that almost all primary teachers, who on the whole taught every subject to their own class, would be required to teach science. This included elements of physics and chemistry as well as the traditional 'nature study' and left many teachers feeling that they possessed insufficient knowledge to deliver the curriculum. This barrier to effective science teaching had been identified some years earlier (DES, 1978) but was still considered to be a difficulty several years after the introduction of the National Curriculum. Russell et al. (1994) found that teachers' 'lack of science understanding inhibited their ability to anticipate the directions in which pupils' learning might proceed'. However, Harlen (1992) reminds us that this is not the whole picture. She argues that 'teaching science to children means enabling them to engage in scientific exploration' and that understanding of this process must accompany possession of factual knowledge.

As the National Curriculum raised the status of science to a core subject and the post of co-ordinator was created for the first time in many schools, there was a need for a good many teachers with a sufficient understanding of science to enable them to support their colleagues. In many cases, the teachers' knowledge of science was some way behind that which they were now expected to pass on to their pupils. There was fear of the new science curriculum, and many teachers were looking to their co-ordinators for more than simple factual knowledge. They required support through the provision of workable activities and guidance on differentiation and progression. There was a clear need in many schools for a subject leader and not just a co-ordinator with a pigeon-hole marked 'science'.

Science, as a specialism in primary initial teacher training courses, had been relatively rare until the 1980s, and clearly some mechanism for the in-service training of the new breed of science curriculum co-ordinators was an important first step in producing the kind of teachers who could respond appropriately to the new curriculum requirements. Two mechanisms were available for this task. For small doses of school-based and day-course Inset, a number of advisory teacher posts were created. To serve the needs of the school co-ordinators there was an extension of what were formerly known as 'designated' courses, supported by the then Department of Education and Science (DES). There was originally no impact on school budgets from teachers' attendance on these courses, as all personnel resource funds were held centrally by local authorities. A teacher's place on one of these courses depended on the teacher's interest, support of head teacher and a crude allocation by local authority advisers. From 1987, funding was provided through the LEA Training Grant System, also known as GRIST (Grant-Related In-Service Training). The advent of locally managed schools with delegated budgets placed responsibility for 'purchase' of training on the head teacher, and the cost of a twenty-day course with its associated supply cover was very difficult to manage, especially for a small school. The situation was ameliorated with the introduction of GEST (Grants for Education Support and Training) which gave local authorities a pool of money which could be credited to schools taking up an allocation of a place on a long course. These courses sought to provide co-ordinators of key subjects with sufficient personal knowledge and management skills to enable them to manage the subject in their school. Assignments were set involving examination of particular aspects of practice in school, and standards were set at a level which allowed teachers to gain accreditation that might eventually lead to the award of a master's degree. Local authority advisers were closely involved in the delivery of these courses, their involvement becoming greater in later years as more of the work became locally focused.

These GEST-funded courses were cut in length between 1995 and 1998 so that many of them lasted only twelve instead of the original twenty days. With the ending of GEST funding of in-service training at the end of academic year 1997–98, institutions were invited to bid for funding for a new range of training which had to meet strict criteria in terms of relevance and effectiveness. When a number of existing providers of science long courses were unsuccessful, many schools felt cut off from their traditional links with higher education, and there has been some reappraisal of the most appropriate way of providing for the career development of teachers. As an example, the Association for Science Education is, at the time of writing, considering ways in which it, as the authoritative voice of science teachers, can contribute to their training beyond its existing provision through local groups and its programme of courses at its annual meeting.

Another trend which might be seen as worrying to teachers intent on advancement of their science education is that of training budgets being increasingly tied to school management or post-inspection action plans. This theoretically gives access to additional training for those whose area of work is identified as being in need of improvement, but may give little opportunity to teachers who have already established a successful career in their subject and wish to take it forward outside the management track. The ideal proposed by Plowden (once separated from the arguments about progressivism and trendiness) of regular challenging and reflective in-service training has been lost, in contrast to many parts of industry, where regular appraisal and training are regarded as essential.

Characteristics of Good Science Teachers

It has been said that good scientists ask good questions. We need to consider here what good science teachers do. We might suggest that good teachers generate such intense curiosity that their pupils cannot fail to ask good questions. However, we live in the real world; one child may ask many really worthwhile questions but not be able to set about investigating any of them, while another may not have the same powers of reasoning but can, through first-hand observation and practical experience, solve problems raised by others. In a classroom situation one meets both of these children, and usually another twenty-eight individuals. So, one characteristic of the good science teacher is the ability to differentiate appropriately based upon knowledge of the child in relation to the specific demands of the subject.

Most teachers will readily describe what Sarah Ford (1998) has called 'the Tingle Factor'. It is that feeling of satisfaction with a job well done when an activity has succeeded in opening up a new area of understanding for a child. Reflective teaching practitioners will ask what it was about the activity that made it successful, and will look beyond the issues of lesson preparation to the pedagogical principles of engaging interest, higher-level questioning, open-ended opportunities and ownership. The question of teacher confidence appears to be central to the pushing out of boundaries beyond the highly structured, closed investigations which characterised science teaching for many generations and are still common today. The purpose here is not to criticise these activities, as they have provided many non-specialist teachers with a way to give children access to their entitlement to a science curriculum, but more to illustrate how a confident teacher, with personal knowledge in reserve, can allow investigations to proceed beyond the conventional and almost into the realms of the creative. Many of our most significant scientific discoveries have come from the minds of those who dared to suggest that frameworks for viewing certain phenomena might exist beyond those conventionally held at the time. Da Vinci took huge creative leaps to suggest that man might fly or ride a bicycle, and his drawings show almost workable devices for both those acts, later 'invented' by others. Galileo's ideas about the universe were regarded as heresy when first proposed, yet were very close to our modern understanding.

It may be useful to attempt to isolate the essential characteristics of science teaching which set it apart from teaching in other areas of the curriculum, and create a list of desirable personal and professional attributes which might characterise a good science teacher. Broad distinctions may be made between those attributes which relate to teachers' knowledge of science, their enthusiasm for it and their ability to communicate it to others. Those rare talented individuals who score highly on all three counts may serve as models for the rest. We are not all blessed with the attributes of those talented communicators we see in the media but we can certainly learn from their communicative techniques.

It is the task of teaching to recognise an individual's potential and develop it. It may also be the task of teacher training to build on the skills which were appropriate for the teaching of science in the 1960s and 1970s and equip teachers for teaching the science of the 1990s and beyond. At the beginning of the 1990s, the National Curriculum Council considered that newly trained teachers should possess 'sufficient subject knowledge to teach and assess pupils across the full range of the National Curriculum levels appropriate to the key stage(s) for which they are being trained' (DES, 1992) (therefore, in the case of a Key Stage 2 teacher, Level 6, which might not seem terribly onerous except that to achieve a Level 6 today involves knowledge not covered in O-level science in the 1960s).

The matter of subject knowledge might at first seem the easiest to address. It might be anticipated that a few evenings with a textbook, backed up by some demonstrations and investigations, would serve to equip serving teachers with a greatly improved understanding of a large volume of subject matter. While this may, in theory, be a workable method, it is a most inefficient one. Learning has been shown to be far more effective when the learner has what Sizmur and Ashby (1997) have termed a 'need to know'. The author questioned a number of colleagues about their experience in acquiring information and communications technology skills and found that most progressed much faster in learning new skills when they had a personal task for the new technique to perform: for example, a need to produce a certain kind of document or a database. Such a 'need to know' motivates an individual through personal commitment to the outcome of that learning. Thus, a skilled trainer of teachers or a skilled teacher of science could do well to develop strategies which create that 'need to know' in the learner. However, as Cochran and Jones (1998) note, even when equipped with what might be considered 'adequate' subject knowledge, a teacher may not be adequately prepared for teaching. Science is continually moving forward, and therefore an updating of knowledge will always be a necessary part of a science teacher's on-going development. Driscoll (1975) rehearses many of the same arguments, noting that in some studies no correlation could be found between levels of subject knowledge and effective teaching. He also points to the fact that many graduate scientists may have full factual knowledge without ever having to engage in genuine scientific enquiry.

To move on to the personal attributes which may characterise a good science teacher, we might attempt to examine what constitutes good teaching. Ofsted (1993) considered that the quality of teaching could be judged by the extent to which:

- teachers have clear objectives for their lessons;
- pupils are aware of these objectives;
- teachers have a secure command of the subject;
- lessons have suitable content;
- activities are well chosen to promote learning of that content;
- activities are presented in ways which will engage and motivate and challenge all pupils, enabling them to make progress at a suitable pace.

This last criterion seems to be the crucial one. If all the first five are in place, then it is the last which is most effective in bringing about real learning. The TTA (1998) specify that trainees should possess pedagogical knowledge and understanding, and should have competence in effective teaching and assessment methods in addition to knowledge and understanding of the subject. Most difficult to develop are the personal skills essential to the effective teaching of science and, in particular, the development of pupils' investigative skills. Some distinct characteristics are required, which may be summarised as below and which link to the previous discussion:

- *Confidence*: in one's subject knowledge; to challenge children's existing frameworks; to set out opportunities for open questioning; to share in the learners' discoveries.

- *Creativity*: in devising strategies which compel the learner to want to know more; in being able to develop models for understanding.

- *Empathy, sensitivity and open-mindedness*: to be able to accept that an alternative framework might just be the right tool for a child to use to make sense of the world as s/he sees it at that time; to judge when is the right moment to intervene; to understand that there are a variety of teaching methods effective in passing on knowledge and skills to the learner.

  When the results of the large body of research into children's alternative frameworks and ways of reconstructing them are considered the picture becomes even more complex. Evidence has been amassed (e.g. Russell and Watt, 1989, the SPACE reports) about the value of using these alternative frameworks as a base from which to build a view on the currently accepted model or scientific explanation of a phenomenon (the constructivist approach – see Chapter 4). However, this presupposes an understanding of the accepted view by the teacher and requires the teacher to be capable of the kind of questioning and interaction necessary to build children's ideas into a conventional framework. Indeed, in the hands of the unskilled, this approach can be counter-productive. Kruger et al. (1990) found a prevalence of certain misconceptions among a sample of primary teachers, and pressed for action to train advisory teachers in science concepts as well as in process skills.

- *Understanding of metacognition*: This phenomenon, described by Flavell et al. (1970), can become a powerful tool in the hands of a reflective practitioner who is prepared to examine children's learning strategies and adopt teaching methods which match them. Nisbet and Shucksmith (1986) advanced the hypothesis that 'children already begin to develop metacognitive knowledge or awareness which could control their strategic activities while they are still in the primary school'.

  To develop pupils' understanding not only of the current body of knowledge but also of the provisional nature of that knowledge and of ways to assimilate new discoveries (personal or external), teachers must have experienced such a process themselves. Driver et. al. (1996) found that when considering the nature of science, secondary school science teachers 'have not had opportunities themselves to reflect on and clarify their own views on the subject'. Hobden and Reiss (1999b) found through the Primary Teacher as Scientist Project (PTAS) that primary teachers' understanding of the nature of science was considerably enhanced by being able to experience for themselves the practice of carrying out scientific research. Many of the teachers involved in the project also became very interested in the consideration of children's alternative frameworks against evidence gathered from direct scientific investigation.

- *Ownership*: Teachers resist pressures to change their practice and will take on what they feel fits their own circumstances, subverting and diluting new practice (Higgins and Leat, 1997). Only when they feel that they have ownership of the curricular philosophy within which they work can they take a completely critical view of their practice. The risk-taking necessary to trial new ideas demands confidence, commitment, ownership and respect from peers. Evidence from the teachers involved in the Primary Teacher as Scientist Project demonstrates that teachers, given ownership of a professional development initiative, have been open to new ways of working in the Science 1 component of the National Curriculum (Sc1) and have passed their confidence to their pupils: 'The pay-off to the children is great. It has put the "wow" back into science for me and for them' (Steve Blakesley, Clwydd).

- *Reflection*: Analysis of the results of small changes in classroom practice (Action Research, or plan–do–review cycles) have become more common among serving teachers. This can be particularly useful for a teacher seeking to make the best use of personal strengths. While Kroath (1989) provides evidence of how such reflection can be guided to enable a teacher to modify subjective theories, Higgins and Leat (1997) indicate that the most useful critical reflection comes with a shift in the locus of control from the managerial to the individual. In many respects we are, once again, considering ownership.

Specialist or Generalist?

The debate about whether to have specialist or generalist teachers has been active for a number of years, but has become more focused as schools have attempted to rationalise their curriculum coverage and, at the same time, make the greatest use of teachers' strengths. It is a matter which can cause some tension for primary teachers torn between their pastoral responsibility as class teachers and a desire to utilise their subject specialism strengths to the full.

Begg et al. (1993), on behalf of ASE, undertook an examination of the issues in the wake of the controversial assertions of Alexander et al. (1992), who, while advocating increased subject specialism, realistically noted that 'The historical funding anomaly between secondary and primary education means that primary schools have insufficient scope to employ, for example, the degree of specialist expertise that is needed to achieve better quality subject teaching.' Begg notes that 'The time for science is particularly difficult to estimate …

[making it] difficult to fit into a rigid timetable, which is a strong argument against subject specialist teachers and rigid timetables in primary schools.'

Ofsted (1997), while finding that specialist teaching of certain subjects has a very beneficial effect on pupils' achievements, caution that children in the classes of a specialist teacher 'may receive teaching of variable quality when they are taught by several other teachers'. My experience adds a further caution – that of the unsettling effect of such movements on certain children for whom the close relationship with a class teacher enables greater progress. However, my analysis, like Ofsted's, finds much of value in a semi-specialist approach. That analysis is reproduced here as a table for ease of reference (see Table 7.1).

The Way Forward

The vision of Plowden influenced a generation and was berated for its lack of structure, yet many of those parts of its message that relate to experiential learning have been borne out by more recent research into the development of children's science process skills. That report's central message stated that 'At the heart of the educational process lies the child.' Its critics might argue that at the heart of the educational process lies the subject. Without an understanding of how the child learns and what triggers retention, our teachers can teach facts till they are blue in the face without success. Those teachers need, themselves, to be fired with enthusiasm for science and its processes.

Another strong recommendation of the Plowden Report was that teachers should receive regular and substantial training post-qualification. This has not happened and is in contrast to industry, where appraisal is often non-threatening and gives access to continuous training with the aim of developing the full potential of every member of staff. Our primary school staff do not have access to such training in the current climate and are frequently at the mercy of the school management plan, which may run several years before bringing their subject to the fore. The bold plans contained in the government's Green Paper (DfEE, 1998) for rewarding excellent teachers might appear to offer incentives to good science teachers to remain in the classroom teaching their specialist subject and spreading good practice to others. However, it is feared that very few classroom teachers of science will benefit from the enhanced opportunities to spread good practice, leaving the majority in a similar position to the present one.

Science is a rapidly developing discipline, and our science teachers should have the opportunity to keep abreast of developments in their subject and of ways of bringing about effective teaching and learning of science. This professional development need not necessarily take the form of conventional instruction but might involve less conventional activities, such as short periods spent in a commercial research environment or working within an institution from another phase of education. Science itself does not stand still, and it thrives on innovative and creative solutions. Science education must learn to do the same.

Table 7.1: Comparison of four models of science teaching

	Advantages	*Disadvantages*
CLASS (GENERALIST) TEACHING (the only model available to many, especially small, primary schools)	Class teacher able to monitor development of whole child Allows small groups to work on a topic while other children are on other tasks Less demand on equipment Science can be integrated in topics where appropriate and taught separately as necessary	Difficult to match task to child Leads to many groups in class and difficulty of interacting with all pupils Unlikely to stretch the most able children Leads to feelings of inadequacy on the part of special needs children
SPECIALIST TEACHING (relies on availability of specialist science teacher)	The person with the greatest science knowledge teaches all the sessions The specialist monitors progress and continuity Other teachers are relieved of responsibility for planning lessons in these areas Specialist gets into other classes and can spread ideas	Class teacher can lose track of progress in science Difficult to teach continuing units of work such as 'weather' The practical nature of much of the work with pupils would mean the specialist spending a large amount of time in preparation Each task must fit the time allocated to it on the timetable
SEMI-SPECIALISM (where a specialist would teach some aspects of science collaborating with established class teachers)	Tasks better matched to children's needs Class teacher better able to monitor development of whole child than with specialist teaching model The specialist monitors progress and continuity Possible to teach continuing units of work such as 'weather'	Children may receive slightly different perceptions from the specialist and class teacher inputs to the same topic
SETTING (by ability or by gender)	Tasks better matched to children's needs Speed of progress can be enhanced	Class teacher can lose track of progress where children not taught by the class teacher Difficult to teach continuing units of work such as 'weather' Equipment and resources insufficient when all sets engaged on activities Science specialist/co-ordinator can teach only one of the sets Timetabling and storage of equipment in use for more than one lesson can be problematic

This chapter has shown that the task of identifying the right kind of teacher for science 3–13 is not an easy one. It depends on many factors: the school, its organisation, curriculum requirements, political influences, developments in the scientific world and, of course, the needs of the children themselves. To return to the opening question, it appears that there are a number of 'right' models which suit different circumstances, and the supply of the right kind of teacher depends on training courses which can prepare a wide range of individuals for the many complex educational situations found in schools.

References

Alexander, A., Rose, J. and Woodhead, C. (1992) *Curriculum Organisation and Classroom Practice*, London: HMSO.

Begg, J., Bell, D., Burton, N., Feasey, R., Goldsworthy, A., Gaskill, G. and Brown, C.M. (1993) *The Whole Curriculum in Primary Schools: Maintaining the Quality in the Teaching of Primary Science*, Hatfield: Association for Science Education.

Cochran, K.F. and Jones, L.L. (1998) 'The subject matter knowledge of preservice science teachers', in B.J. Fraser and K.G. Tobin (eds) *International Handbook of Science Education*, London: Taylor & Francis, pp. 707–18.

DES (1978) *Primary Education in England*, London: HMSO.

DES (1989) *Science in the National Curriculum*, London: HMSO.

DES (1992) *Science, Key Stages 1, 2, and 3. A Report by HM Inspectorate on the Second Year, 1990–91*, London: HMSO.

DfEE (1998) *Teachers Meeting the Challenge of Change*, London: The Stationery Office.

Driscoll, D. (1975) 'Trends in education of science teachers', in P.L. Gardner (ed.) *The Structure of Science Education*, Victoria, Australia: Longman.

Driver, R., Leach, J., Millar, R. and Scott, P. (1996) *Young People's Images of Science*, Buckingham: Open University Press.

Flavell, J.H., Friedrichs, A.G. and Hoyt, J.D. (1970) 'Developmental changes in memorization processes', *Cognitive Psychology*, 1: 324–40.

Ford, S. (1998) 'Bring back the Tingle Factor', *Education in Science*, 177: 3.

Goldsmith, S. (Subject Officer, QCA) (1998) Address on current issues in science education, given to The Education Show, NEC, Birmingham, March.

Harlen, W. (1992) *The Teaching of Science*, London: David Fulton.

Higgins, S. and Leat, D. (1997) 'Horses for courses or courses for horses: what is effective teacher development?', *British Journal of In-service Education*, 23 (3): 303–15.

Hobden, J. and Reiss, M. (1999a) *The Primary Teacher as Scientist Project*, Leicester: National Centre for Initial Teacher Training in Primary Science.

Hobden, J. and Reiss, M. (1999b) Editorial, *Primary Science Review*, 56 (Special Issue): 2–3.

Hudson, J. and Slack, D. (1981) *Science Horizons – The West Sussex Science 5–14 Scheme*, Basingstoke: Globe Education.

Jackson, J. (1998) 'Improving science education 5–14', unpublished keynote address to the Annual Meeting of the Association for Science Education (Scotland), 7 March.

Jarman, R. (1993) 'Real experiments with Bunsen burners', *School Science Review*, 74: 19–29.

Kroath, F. (1989) 'How do teachers change their practical theories?', *Cambridge Journal of Education*, 19 (1): 58–69.

Kruger, C., Summers, M. and Palacio, D. (1990) 'Inset for Primary Science in the National Curriculum in England and Wales. Are the real needs of teachers perceived?', *Journal of Education for Teaching*, 16 (2): 133–46.

National Curriculum Council (1993) *Teaching Science at Key Stages 1 and 2*, York: NCC.

Nisbet, J. and Shucksmith, J. (1986) *Learning Strategies*, London: Routledge.

Ofsted (1993) *Handbook for the Inspection of Schools*, London: HMSO.

Ofsted (1997) *Using Subject Specialists to Promote High Standards at Key Stage 2*, London: HMSO.

Pattinson, R. (1998) 'The nature of scientific investigation', *Primary Science Review*, 53: 25.

Ponchaud, R. (1998) 'Quality in science education' in *ASE Guide to Primary Science Education*, Hatfield: ASE.

Russell, T. and Watt, D. (1989) *Primary SPACE Report: Growth*, Liverpool: Liverpool University Press.

Russell, T., Qualter, R. and McGuigan, L. (1994) *Evaluation of the Implementation of Science in the National Curriculum*, London: SCAA.

Sizmur, S. and Ashby, J. (1997) *Introducing Scientific Concepts to Children*, Slough: NFER.

TTA (1998) *Initial Teacher Training National Curriculum for Primary Science* (annex E of DfEE Circular 4/98), London: TTA.

West Sussex County Council (1993) *New Horizons: Science*, Cambridge: Cambridge University Press.

Wolpert, L. (1992) *The Unnatural Nature of Science*, London: Faber.

8 The Impact of Information Technology

Angela McFarlane

When considering the relationship between information technology (IT) and science in primary schools it is impossible to separate this from the more general history of computers in schools. Despite the introduction of computers into UK schools in the early 1980s, their presence has failed to transform the curriculum, or school culture. This is not because the technology lacks the potential to act as an agent of change, but because technology alone will not transform schools from the evolutionary organisations they certainly are into revolutionary ones. And yet it is not that schools have remained unchanged during the first two decades of the information age. Indeed, primary schools in particular have undergone transition from a child-centred, integrated curriculum widespread in the 1970s, to a whole-class, subject-oriented approach mediated through statute and policed through external assessment and inspection. This chapter will consider the story of computers in primary schools since 1978, with a focus on science. In particular, it will attempt to identify the factors that have inhibited widespread use of IT in primary science, and give an indication of what the role of IT in science could be, both now and in the future.

1978–1989

The year 1978 could be seen as the birth date of the microcomputer – certainly the first products exploiting the chip architecture were seen then. The first initiative to put computers in schools was not until 1981, and it was funded not by the ministry responsible for education but by that for trade and industry (Scrimshaw and Boyd-Barrett, 1991). The earliest computer to find a widespread home in primary schools, the Acorn BBC, was widely regarded as a 'science-friendly' computer. There were two main reasons for this; the first was that it had lots of 'ports' in the back where various probes and sensors could be attached, the second that it could display simple coloured animations which could be influenced by the user. As a result of these two features, there were a number of publications in school science aimed at either developing the use of the BBC for what became known as 'data logging', or using the computer to run a model or simulation (Sparkes, 1984; Brankin and Dunkerton, 1986). However, these publications confined themselves to the secondary sector, probably for two reasons. At that time, in the early 1980s, computers were rare in primary schools, with one per school still a government objective in this phase. Investigative science was not yet part of the *statutory* primary curriculum, so there was little science curriculum demand for increasing this objective. There were also few people involved in curriculum innovation with experience and expertise in the three domains of IT, science and the primary phase at that time.

The first government initiative to explore the role of the computer in learning, and including significant funding for software development, was the Microelectronics Education Programme (MEP, 1981–6). As a result of that activity, and the associated flurry of interest it both reflected and created, a wide range of programs of educational value were created. Some of these were aimed at science, and a small number did find a place in the primary curriculum. The most notable of these must be 'Branch', a program that created dichotomous keys using a series of questions entered by the user. The answer to each question was yes or no. Keys created in this way could then be used to identify one of the set of objects for which the key was built. This program fitted very well into a topic on sorting. 'Branch' represents a genre of computer-aided learning (CAL) software which was popular and produced in profusion at that time. Programming the BBC was relatively easy; small but powerful programs could be produced by enthusiastic amateurs who had an imaginative teaching method and could turn it into an interactive learning tool. The idea of teachers programming software in their back bedroom was still a realistic possibility, and some educational software companies which started in this way are still in existence some twenty years on.

How widespread the use of such software was in primary science during the MEP is not directly recorded. However, the low level of hardware access, and the relatively low profile of science in the primary curriculum, suggests this was far from being a mainstream activity in primary schools in the UK (see Table 8.1).

Table 8.1: Figures for the primary sector in England and Wales

Year	*Pupils/Computer*	*Computers/School*	*% of schools using IT in science*
84/5	107	1.7	Not available
88/9	67	2.5	Not available
94/5	18	9.9	76
96/7	19	13	4 substantial 47 some 44 little 4 none
97/8	18	13	4 substantial 52 some 38 little 6 none

Source: From the statistical surveys of IT in schools published in 1985, 1989, 1995 and 1998 by the DES and DfEE. Based on a sample of 1,000+ schools.

Even in the secondary sector, the use of computers to support science was a 'hobbyist' activity until the late 1980s. There were pioneering teachers and innovators who explored the possibilities of interfacing sensors to the BBC and using the machine to record and display measurements (this activity was then known by a number of names, including 'interfacing', but is now known as data logging). These people came mostly from a physics background and often designed and made their own boxes of electronics to connect sensors, such as light gates and thermistors, to the computer. The first such commercial device to find a significant place in schools was the VELA. This was an expensive and complex, albeit versatile, device which attracted a small but significant following in secondary schools. Yet it was not until the first government-funded push to support IT in schools through the Education Support Grant (ESG) in 1988, when identified advisory teachers for IT with strong subject backgrounds were funded, that the role of computers in primary science began to be investigated seriously. The government agency, then known as the Microelectronics Education Support Unit (MESU) – a name which captures the technocentric perspective of the time – had an officer who was particularly interested in primary science. He supported the development and use of a simple interface and sensors with software designed for the primary user. This support continued when MESU was replaced by the National Council for Educational Technology (NCET) in late 1988. As the profile of data logging in primary schools continued to rise, a wide and bewildering range of devices became available, each with its own software (McFarlane, 1991, 1992). NCET withdrew from directly supporting the development of hardware or software for data logging, leaving these to the commercial sector. This left the government agency free to concentrate on the production of curriculum support material, with some purchasing information (NCET, 1994). They also worked with the Schools Curriculum and Assessment Agency (SCAA), now the Qualifications and Curriculum Agency (QCA), to produce non-statutory guidance on the use of IT in science. At the end of the 1990s these were still functions of the British Education and Communications Technology Agency (BECTa), the organisation that replaced NCET.

It was not until the National Curriculum was first established in 1989 that investigative science found a home, at least in theory, in all primary schools. Moreover, there was an entire Attainment Target dealing with microelectronics, but the real emphasis here was on the microelectronics as an object of study, not as a tool to promote understanding in other areas of the science curriculum. The content, as a result, had little relevance to the primary school. So the 1980s ended with primary schools ill equipped to use IT in teaching, and with no statutory obligation to do so.

1990–1995

The DfEE statistics collected during the early 1990s suggest that while the level of access to machines rose slowly but steadily, as did teacher confidence, the level of overall use of IT in the primary classroom did not keep pace. Teachers were not embracing IT. In an attempt to provide evidence to inform practice and show how the use of IT can enhance learning, the DfEE commissioned what became the ImpacT report, published in 1993 (Watson, 1993). This looked at the use of IT in primary and secondary maths, English and science (as well as secondary geography). Rather than identify schools where significant use of IT was established as its experimental base, this project chose a randomised sample of schools and then tried to identify pairs of classes which were high or low users of IT. These classes were identified from the level of intended use over the coming year. The type of IT used in primary science included databases, spreadsheets and word processors with only one school in the sample of nine using any kind of CAL science software. The project used pre- and post-test comparisons of 'general intellectual enhancement' (Watson, 1993, p. 15) using modified versions of the Assessment of Performance Unit (APU) tests to gauge scientific reasoning. The project ran for two years for primary science, at the end of which the report concluded that 'The case for IT in science was not supported – with a significant difference in favour of the LoIT group' (p. 2).

This finding was reported in the main report and the summary report, and was picked up in the national press. The report also showed that 'IT use in science was generally low when compared with the use in other ImpacT HiIT classes' (p. 3). It was only if the actual data was examined in the main report that it became clear that the finding as reported was highly misleading. The sample of classes was small; three for HiIT and four for LoIT. One HiIT class had only 4 hours' use per term, the same as one of the LoIT classes. Another LoIT class had had 30 hours in the same period, almost twice as much as any HiIT class. Some designated low IT users at the beginning of the project had in fact used much more IT during the period of study than those designated high IT users (p. 32) . The labels HiIT and LoIT were retained, giving the impression that IT use in primary science led to a poorer performance. This highly damaging interpretation was widespread and very nearly led to the exclusion of primary science from the next round of targeted government funding for IT.

The year 1993 saw the first of a series of portable computer pilots run by NCET. These projects put thousands of powerful portable computers into schools in over a hundred separate curriculum-oriented projects, and evaluated the outcomes in a range of subjects, with an emphasis on mathematics, English and science. One project, directed by the current author, looked at the impact of such machines, with data-logging equipment, on the development of both investigative and graphing skills in science. The outcomes showed clearly that the ability of pupils to grasp complex skills they otherwise found difficult, if not impossible, at that age, was enormously accelerated after relatively short exposure to data logging (Warwick and McFarlane, 1994; McFarlane, 1994; McFarlane et al., 1995). This was the first hard evidence of enhanced learning in live classroom situations associated with computer use in primary science in the UK.

The computers being purchased by and for schools at this time were a very different animal from those of the early 1980s. Gone were the lumpy four-colour graphics and the use of tape drives. Machines now had user interfaces resembling those of today, with words and pictures on screen to represent powerful command sets. Moving a pointer and clicking a button replaced incomprehensible typed commands. Mouse-driven software had arrived. Another more significant development also took place at this time, that of the change of emphasis from curriculum-specific software to the use of generic-application software. Personal computers had become the major commercial business they had been predicted, and continue, to be. The major market which shaped their development was commerce, not education. Powerful software to manipulate digital data in the form of words and numbers emerged. Education was quick to exploit the power of word processors, spreadsheets and databases but it must be remembered that these packages were never designed for education and still lacked key features. The most obvious example for science is that early spreadsheets did not support x–y plotting in the graphing features. In fact, this remained highly cumbersome in major packages until very recently.

At the same time as the computer itself became easier to use, it became harder to program. Programming software for the Microsoft Windows environment in the early 1990s was akin to a black art, strictly for larger commercial concerns. Certainly it was beyond the back-bedroom programmer who could not have the time or resources to develop even small programs. Perhaps as a result, the number of science CAL packages for these newer machines was, and has remained, small. As yet there is still no equivalent to the Branch program, despite its earlier popularity. As educational software development became a largely commercial activity, the small market for such subject-specific software compared to costs of versioning for the new machines, let alone new development, was not economically viable.

There was a drive to encourage use of word processors, spreadsheets and databases across the curriculum. CAL dropped from view. Science, though, did have one important difference. Data logging and control had entered the National Curriculum. It was this application of IT, specific to science in the case of data logging, which now received the full emphasis of curriculum support offered to teachers of science in both the primary and the secondary schools. Control never really found a home in science and remained the preserve of the technology curriculum.

1995

The revision of the National Curriculum in 1995 gave a much higher profile to IT in science in all phases, including its use to support learning 'where appropriate'. However, use still remained non-statutory. (Statutory inclusion of IT would have left the government ultimately liable for the resource and training implications that this would have entailed.) This was a key year as far as IT in primary science was concerned. For the first time there was a statutory curriculum that referred to the use of data logging in Key Stage 2 as well as more general, if vague, encouragement for wider use of IT. There was a range of equipment and software developed for the sector, copious free printed support materials (e.g. Frost et al., 1994; McFarlane, 1993) through agencies such as NCET and support staff in most LEAs. The scene was surely set for a boom. Yet there were still three highly significant issues unaddressed: the level of access to computers in primary schools, the level of competence of teachers to use IT generally, and the fact that teachers did not understand the synergy between the use of IT and learning in science. This in turn was not reflected in the support materials, which emphasised how to use IT, not why, and the end of Key Stage statutory assessments ('SATs') in science for Key Stages 1 and 2 certainly made no reference to IT or relied on any experience of its use. The numbers of schools reporting the use of data logging or control stayed well below 10 per cent (DfEE, 1995). The uses of CAL software, databases and spreadsheets – the software applications most likely to be relevant directly to science – remained very low; the only significant use of computers in primary classrooms was for word processing. The boom did not come.

The Situation at the End of the 1990s

In science as practised and taught outside school the use of computers and computerised equipment is ubiquitous. As long ago as 1991 the *Primary Science Review*, in a special edition devoted to IT in science, gave a comprehensive overview of the role of IT in primary science (Ovens, 1991). This was further developed in the 'Enhancing Science with IT' materials in 1994 (Frost et al.) (See McFarlane, 1997, for a more detailed discussion of the role of ICT in primary learning, including science and related mathematical skills in particular.) See Figure 8.1.

Eight years later, we might have expected to find these uses embedded in the primary science curriculum. And yet DfEE statistics suggest that the use of computers in school science has remained a 'Cinderella subject'. In 1997/8 total average use of computers by pupils in primary schools amounted to one session per week, with most of this use reported in English and maths. Although 92 per cent of schools reported use of IT in primary science, only 4 per cent reported substantial use (see Table 8.1). What they were using is not recorded, but levels of use of data measurement and control, and simulations, remained very low.

Clearly there are other priorities at the top of the primary school agenda at the end of the 1990s. It is interesting that after a rapid increase in the number of computers in primary schools in the early 1990s, the numbers have stabilised since 1995 at around one per eighteen pupils. The literacy and numeracy strategies and their implementation to meet targets for SATs in maths and English in 2002 relegates other subjects to second place. Furthermore, the lack of use of IT within the strategies justifies the use of less rather than more IT in primary schools. The importance of science, let alone IT in science, has perhaps lessened.

One explanation for the failure of IT to take its place at the heart of investigative primary science is that teachers have yet to be convinced of its

Figure 8.1: Integrating IT into AT1

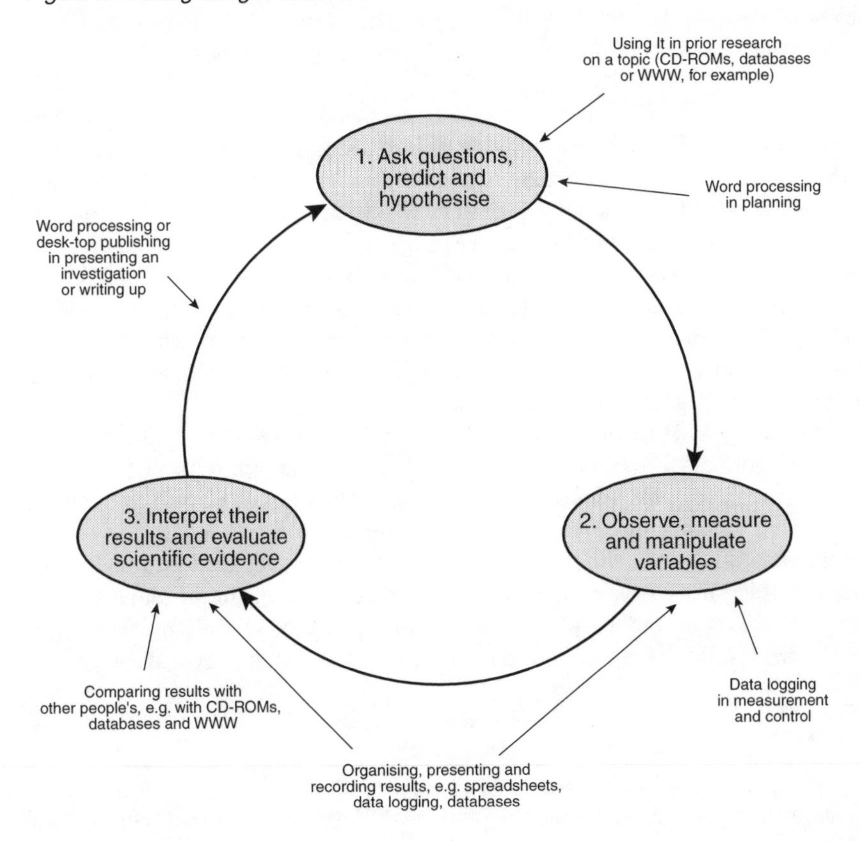

importance or its potential. This is hardly surprising given the lack of relevant training that teachers have had, and the fact that the training and support materials offered have concentrated on what can be done and how, rather than why. Ninety per cent of teachers have received some training in the use of IT, but little of this is substantial – i.e. more than an introductory course which was unlikely to have looked at science specifically.

Available support materials may have reference to Attainment Targets, but no clear relationship to specific teaching objectives or how these can be better achieved through the use of IT. Data logging in particular is a complex use of IT, involving as it does additional plug-ins to the computer, and specialist knowledge of investigative science and data analysis as well as IT capability on the part of the teacher. A teacher needs to be a confident manager of IT in the classroom and have clearly identified additional value to pupils before she is likely to want the hassle. An obvious identified additional value, and one which increasingly orders priorities in schools, is external recognition. This might come in the form of an Ofsted inspection or SATs results. If teachers knew that Ofsted inspectors would look specifically for the integration of IT in science lessons, or that science SAT tests would assume use of IT and have questions which could not be tackled as well (or at all) unless the pupils had used IT in a science context, the situation might be different. As it is, neither of these things is likely.

Even where an enthusiastic IT or science co-ordinator, or both, might recognise the value of IT in meeting science objectives and wish to introduce more use, there are significant resourcing issues. Within a primary school budget, the cost of specialist IT or science equipment is difficult to find. It is hard to get data on actual use of IT, but recent research suggests that most primary pupils use a computer once a week in school, with some getting to use one up to twice a day, and others only once a fortnight. Given that this is total use, not use in science, it is important to remember that the reality in most schools is that access to computers is minimal – and that it has to address the whole of the requirement for IT across the curriculum.

It is not that good equipment and models for its use do not exist. There is now a range of products for science based around the cheaper, more robust portable computers such as palmtops or the specialist educational portables such as DreamWriter. These include equipment and software for data logging as well as the generic applications, and even a modem for Internet and e-mail access. Moreover, this kind of equipment has proved its worth in the primary science context (McFarlane and Freidler, 1998). Some primary schools recognise the power of these kinds of resource and find the funds. Most, however, do not.

The Future

The number of children with access to IT outside the school grows steadily and there is no reason to expect that this will stop. Between 35 and 80 per cent of children have home access now. Government is concerned for the disadvantaged, the 'information poor', who do not have home provision. As a result there is major investment in public libraries, now set to be exciting and accessible centres providing access to the Internet for all. By the end of 1999 the seventy libraries in Essex, for example, will have computers with on-line access freely available. Smaller libraries will have ten, others will have forty. Primary schools cannot match this. Currently 7 per cent have on-line access, and that is usually one line into the office used for data swapping with the LEA. Curriculum use is negligible. The planned National Grid for Learning (NGfL) related spend of £750 million on infrastructure is set to ensure that all primary schools will have this level of on-line access. The moneys will not stretch to networking classrooms, however, and curriculum use is predicted to remain low as a result.

The library-based learning scenarios of the future, described in the Library and Information Commission Report, *New Library: The People's Network* (1997) should give teachers pause for thought:

> Zahir is five years old, and his grandmother brings him to the local library. When he has read a book he really enjoys, he races to the computer to

> write what he thinks of it. Anyone else looking up that book can see what he has written. His younger sister speaks her views of the picture-books she likes, and the computer makes her words appear on the screen.
>
> There is a special game he plays with his two friends in which they can use the computer to make up a story of their own with different endings and then print it out. If they get stuck on a word, the dictionary helps – sometimes with a picture or a moving image of what they are trying to describe.
>
> When he brings his books back, the librarian suggests other books he might enjoy, as the library has a list of all the ones he has borrowed before and written something good about. He prefers to choose his own, though, and has enjoyed trying to read a really long book on dinosaurs. When he brings it back on Sunday afternoon, the lady gives him a list of stories about dinosaurs – he chooses the ones about Dilly the Dinosaur, as someone his age who lives in the Punjab thought it was very good – and she also tells his grandmother about the Dinamites exhibition that is on this week. His grandmother promises to take him there, and tells him he can visit it again on the CD-ROM in the library.
>
> (p. 4)

And yet ICT (information and communications technology), especially the Internet, offers new and unique opportunities to the primary science teacher. There are chances to communicate with science experts such as vulcanologists, get updates on the latest space probe, see pictures of distant planets as they are captured, find out about the weather in Australia. Many museums have sites with information which can be used to prepare for or follow up visits. Popular science programmes on television and radio offer sites with information especially for children, even where the original programme might have been for an adult audience. If schools are not careful, the experience of science they offer pupils will become dull and uninspiring compared to that offered by the local library or the science television channel.

In science, as with the rest of the curriculum, schools must be able to embrace new technologies. Teachers must become facilitators guiding children through the exciting range of information and experiences available to them. In order for this to happen, however, there are issues of training, resources and policy to be addressed. Teachers need to develop their own technological literacy, and to understand how to become effective facilitators. Moreover, teachers and children need better access. The Multimedia Portables for Teachers project showed how powerful personal ownership of the hardware and on-line access is as an agent for professional development. Ninety-eight per cent of the teachers involved in the first phase improved and increased their use of ICT in the classroom (NCET, 1998). This result was so outstanding that in the following year, 1998, the government spent a further £23 million on an extension to the project. In total some 10,500 computers were provided for teachers in 5,000 schools. Disappointingly, the £230 million to be spent on in-service training for teachers and librarians in the years leading to the year 2002 cannot be spent on subsidising a personal purchase of a portable computer.

Giving pupils personal access to portable computers has also proved a powerful agent for enhanced learning (Stradling et al., 1994; McFarlane and Freidler, 1998). Schools will never be able to provide access at point of need for all pupils if they have to provide the computers. However, within ten years the price of a suitable portable should reach that of a good bicycle. That will at last open up the possibility for most children to come to school with a machine, as they do now with a calculator or even pencil case. Schools will provide equipment for those who cannot afford a computer, as they do school meals. The school resource budget will then be free to provide the additional infrastructure, such as broadband access to on-line media, good scanning and printing facilities and access to specialist science software and equipment, rather than base machines, which is where current spending is concentrated.

Policy changes may prove harder and take longer. Current emphasis on spelling, grammar and counting, improved through whole-class teaching, is at the other end of a spectrum of models of learning from the child-centred model depicted in the example taken from the library of the future above. Existing policy, and assessment structures based on it, values and rewards the knowledge of a specified body of content. There is little recognition of the skills needed to analyse, interpret and synthesise personal knowledge from a range of experiences and sources. Neither is there an opportunity to foster these skills in an individual child through her excitement in and prior knowledge of a topic. It does not matter how easy it would be to construct a personal learning plan for a child thrilled by dinosaurs, through which key skills in science would be explored and developed – dinosaurs are not on the curriculum, and anyway, not all the class wants to do them.

In a drive to raise standards, primary schools are being forced to abandon the child-centred approach which had been well established in our schools. This makes it harder to harness children's natural enthusiasm for learning. In science we have the advantage still of an exciting subject, with an important place for the development of investigative skills which will have an even wider relevance. We are in danger, however, of being overtaken by the more authentic experiences of science available to children outside school. Schools must harness the technologies which provide that access and teachers must act as guides to children's learning as they navigate this sea of meaning. This cannot happen unless policy-makers too play their part.

References

Brankin, C. and Dunkerton, J. (1986) *Science On-Line – Eye-catching Science Experiments for the BBC Micro*, London: Hodder and Stoughton.

DES (1985) *Survey of Information Technology in Schools, Statistical Bulletin*, London: Department of Education and Science.

DES (1989) *Survey of Information Technology in Schools, Statistical Bulletin*, London: Department of Education and Science.

DfEE (1995) *Survey of Information Technology in Schools, Statistical Bulletin*, London: Department for Education and Employment.

DfEE (1999) *Survey of Information Technology in Schools, Statistical Bulletin*, London: Department for Education and Employment.

Frost, R., McFarlane, A.E., Hemsley, K., Wardle, J. and Wellington, J. (1994) *Enhancing Science with IT*, Coventry: National Council for Educational Technology.

Library and Information Commission (1997) *New Library: The People's Network*, HYPERLINK: [http://www.ukoln.ac.uk/services/lic/newlibrary/]

McFarlane, A. E. (1991) *Technical Bulletin. Resources for Data Monitoring and Control for Science Education*, Coventry: NCET.

McFarlane, A. E. (1992) *Technical Bulletin. Resources for Data Monitoring and Control for Science Education*, Coventry: NCET.

McFarlane, A. E. (1993) *Data Logging in Science* (video and booklet), Coventry: NCET.

McFarlane, A. E. (1994) 'Using computers to effect learning gains in science', in *Proceedings of the Tenth Annual Conference of Informatics in Schools*, Polish Ministry of Education.

McFarlane, A. E. (ed.) (1997) *IT and Authentic Learning – Realising the Potential of Computers in the Primary Classroom*, London: Routledge.

McFarlane, A. E. and Freidler, Y. (1998) 'Where you want IT when you want IT – the role of portable computers in science education' in B. Fraser and K. Tobin (eds) *International Handbook of Science Education*, Amsterdam: Kluwer.

McFarlane, A. E., Freidler, Y., Warwick, P. and Chaplain, R. (1995) 'Developing an understanding of the meaning of line graphs in primary science investigations, using portable computers and data logging software', *Journal of Computers in Mathematics and Science Teaching*, 14 (4): 461–80.

NCET (1994) *Primary Science Investigations with IT, Using IT to support Key Stage 2 Science*, Coventry: NCET.

NCET (1998) *Portable Multimedia Computers: IT Skills for Teachers*, Coventry: NCET.

Ovens, P. (ed.) (1991) *Primary Science Review 20*, Hatfield: The Association for Science Education.

Scrimshaw, P. and Boyd-Barrett, O. (1991) *Technologies for Teaching; The Use of Technologies for Teaching and Learning in Primary and Secondary Schools*, London: Parliamentary Office of Science and Technology (POST).

Sparkes, R. A. (1984) *The BBC Microcomputer in Science Teaching*, London: Hutchinson.

Stradling, B., Sims, D. and Jamieson, J. (1994) *Portable Computers Pilot Evaluation Report*, Coventry: NCET.

Warwick, P. and McFarlane, A. E. (1994) 'IT in primary investigations', *Primary Science Review*, 36: 22–5.

Watson, D. M. (ed.) (1993) *The ImpacT Report, an Evaluation of the Impact of Information Technology on Children's Achievements in Primary and Secondary Schools*, London: DfEE and King's College, London.

9 The Influence of Interest Groups on 3–13 Science

Philip Stephenson

This chapter will examine the range of non-governmental agencies which, for a variety of reasons, seek to influence the content and quality of science education in the primary school. It will reflect on the changes in the nature of non-governmental support for primary science over the previous fifteen years, will consider the current role that each of the various categories of such support plays, and will then examine how this support and influence is likely to progress in the future. Consideration will be given to implementation and evaluation in terms of how the teaching profession has responded to these initiatives of the interest groups. Whatever the rationale of any specific interest group, it will become clear that the degree of success in terms of affecting the quality of science education in primary schools is dependent on many factors. These include how aware teachers are of the support that non-governmental interest groups offer, whether teachers perceive there to be an 'ulterior motive' and, most significantly, whether the support maps on to the highly defined curriculum structures that the majority of primary schools currently operate. These vital implementation issues are examined in the light of research conducted in a wide sample of schools.

The interest groups considered here can be broadly classified as professional organisations such as the Association for Science Education (ASE), charitable trusts including the Nuffield Foundation, promoters of the public understanding of science such as the British Association (previously the British Association for the Advancement of Science, and still abbreviated BAAS), 'single focus' charitable organisations represented by the Royal Society for the Prevention of Cruelty to Animals (RSPCA), and industrial/commercial groups which include individual corporations such as Glaxo Wellcome and collective professional bodies such as the Society of Chemical Industry (SCI). The involvement of this last group of organisations in science education is often co-ordinated by industrial liaison groups such as the Science and Technology Regional Organisations (SATRO) and the Chemical Industry Education Centre (CIEC). Equally, industry and commerce will frequently commission educationalists such as local education authority (LEA) advisory teachers and initial teacher trainers to facilitate the introduction of products and initiatives for primary schools and thereby lend them an air of credibility.

It is important to understand the context in which these interest groups have operated. Prior to the 1988 Education Reform Act and the introduction of the National Curriculum, science in primary schools was supported in the main by LEA inspectors and advisory groups. These provided what were, by today's standards, somewhat slim recommended guidelines, which had varying degrees of success mainly determined by the level of priority given to science by the individual school. In these pre-National Curriculum years there was, however, increasing recognition of the importance of children engaging in scientific activity and considering scientific ideas.

The Science 5–13 Project 1967–74 (sponsored jointly by the Schools Council, the Nuffield Foundation and the Scottish Education Department) can be clearly seen as a basis upon which the Key Stage 1/2 elements of the National Curriculum were constructed some fifteen years later. The outcome of this project was the series of *Macdonald Science 5–13* books, which were widely taken up by primary schools during the 1970s. It is interesting to note the involvement of the charitable Nuffield Foundation in co-sponsoring the project. Nuffield had a strong track record regarding secondary science, but this was the first real recognition of the importance of the primary phase. It is also important to recognise that many of those involved in the project were ASE members and that they, alongside other interested colleagues, were instrumental in raising the profile of primary science in the previously secondary-dominated world of the ASE. The clear influence that this project had on both the content and, more important perhaps, the investigative approach and associated attitudes expressed in the National Curriculum (see Chapters 2 and 5) indicates the importance and influence of non-governmental interest groups on the development of primary science. As will be seen, ASE and Nuffield were not alone in effecting and bolstering change, but the significance of their role as pro-active innovators makes them a useful starting point in analysing how and why non-governmental influence operates. It will be useful, therefore, to start with an overview of the work and intentions of these and other groups.

Professional Associations: The Association for Science Education

Under the banner 'Teachers helping teachers to teach science', the ASE is an organisation of over 24,000 teachers, advisers, technicians, industrialists and others contributing to science education. It is a registered charity, financed by members' contributions, and receives no government funding. The ASE offers support, advice and information to individual teachers, schools and colleges. In doing so it promotes, supports and develops science education, from primary through to tertiary levels, and informs those involved in industry and commerce. There is a collaborative ethos that offers support, advice and information to other societies and associations with similar aims. The ASE provides a forum for the views of members on science education issues through its regional and national committee structures. It provides many links with industry – arranging teacher secondments and visits to industrial sites – and promotes relevant curriculum-support materials. Most importantly, in terms of influence, the ASE is frequently and regularly consulted by those in authority: government, industry and LEAs. It is interesting to note that the membership of government consultation and working groups, developing both the science National Curriculum and the initial teacher training curriculum, is dominated by ASE members.

Charitable Foundations: The Nuffield Foundation

Founded in 1943, this most prolific of philanthropic organisations resolved to take bold initiatives, working on the principle of setting its own agenda rather than simply reacting to proposals which came from outside. Since its first consideration of science teaching in the early 1960s, this leadership role has been maintained.

The development of the Science 5–13 and Secondary Science modules were characterised by the innovative approach to teaching set in the context of progressive theories of learning. The Nuffield 'approach', based on the principles of guided discovery, represented a paradigm shift in terms of the way educators thought about science. This innovative approach has been maintained with the most recent initiative resulting from the Nuffield Foundation, the development of the Nuffield Primary Science Project. Work on the project continued throughout the 1990s and was based on the findings of the Science Processes and Concepts Explorations (SPACE) research conducted at Liverpool University (see Chapters 4 and 5).

Throughout the life of the Nuffield Primary Science Project, a programme of dissemination has been maintained. This includes a newsletter; talks and workshops for teachers, trainee teachers and advisers; a network of advisory teachers and teacher educators from higher education, including members of the original SPACE team; and collaboration with other Nuffield projects, most notably Nuffield Exeter Extending Literacy (EXEL). The resulting scheme for schools, including 'big books' aimed at integrating science into the literacy hour initiative, has been crucial in encouraging an approach to teaching that reflects an understanding of the constructivist theory of learning.

Organisations Promoting the Public Understanding of Science: The British Association

During the 1970s, the BAAS recognised the shifting emphasis towards the primary sector, which had come about largely through the influence of the Plowden Report. The BAAS, now a registered charity, was founded as the British Association for the Advancement of Science in 1831 by members of 'educated society' (clergy and those involved in science). It provided a forum for discussion and dissemination; it was at the 1868 conference that those attending were called 'scientists', the first use of the word. Today it promotes science and technology in a variety of ways including the support of educators. Funding is essentially derived from industry, which is increasingly happy to give the BAAS education specialists a free hand in what they produce for the market, as long as the sponsoring concern is recognised. Other sources of money include charitable foundations such as the Gatsby Charitable Foundation and Wellcome Trust. There is some government support but, as is the case with commercial and charitable support, for specific projects only. Wherever the money comes from, the BAAS runs on short-term project funding.

Interest from external groups began in earnest in the late 1970s. Industry and charitable foundations had been supporting secondary for some time. However, support for primary science was minimal. This was based on the perception that what science was going on in primary schools was sporadic and relied on whether the school had an 'enthusiast', together with the perception that there appeared to be little external curriculum pressure to teach science. A second reason for potential sponsors' lack of interest in supporting primary science, particularly by industry, was that it was not viewed as being effective in terms of future recruitment which was, at the time, the main funding incentive. It was in this climate of progress and change that the BAAS launched British Association Youth Section (BAYS) national network of science clubs. Membership was taken up essentially by schools and youth groups. Although the rationale was to provide materials to support science activity in after-school clubs, it was natural that teachers would start to integrate the materials into their classroom teaching. In fact, for many teachers, this was their first introduction to an 'investigative' approach. Many of the activities that were novel at the time appear in numerous schemes and text books today.

Due to the timing of the introduction of BAYS, it is difficult to see if it had direct influence on the content of the National Curriculum – a tempting notion, particularly with regard to Sc1. It is more likely that the BAYS materials reflected a general change in terms of attitudes and policy towards primary science. It must be noted that the BAYS materials were written partly by teachers and partly by LEA advisory teams, who were being equally influential in advising the National Curriculum Council in their deliberations about what would go into the 1988 National Curriculum for science. What is in no doubt is the influence that the BAYS initiative has had over the past fifteen years on primary and middle school science, particularly in those subscribing schools (between 1,250 and 1,300 at any one time: potentially 15,000 children, even at a conservative estimate). Sadly, there is no formal evaluation, bar the fact that schools continue to sign up. While teachers have informally reported that those participating in after-school club activities show an understandable increased confidence and ability in the classroom, there is, as yet, no concrete research evidence to support this assertion. Some might argue that the type of child who would participate in an after-school science club is the type of child who would perform well anyway.

Single-issue Organisations: Royal Society for the Prevention of Cruelty to Animals (RSPCA)

The RSPCA was established to improve the human condition by reducing the level of cruelty around people. Remarkably, the National Society for the

Prevention of Cruelty to Children (NSPCC) evolved out of the RSPCA, as a result of animal welfare officers entering properties to deal with animal cruelty, only to witness cruelty to children as well. From the outset, there was a clear educative role and a specific action point of particular vision – to prepare materials for use in schools. In the early 1960s the education department began to recruit educationalists to serve their needs. The purpose was the promotion of animal welfare through education, thereby meeting one of the RSPCA aims as a charitable organisation. The work is based on a belief that behavioural changes create and spring from changes in attitude. Therefore, the organisation sets out to raise awareness and to develop respect and responsibility. Throughout the 1970s, there were two main issues addressed in their education policy:

- The opposition to dissection (essentially in secondary schools) where animals were used without any systematic attempt to present related ethical issues, with animals perceived as 'tools', not sentient beings.
- The improvement in the keeping of animals in the education system (with a clear primary-school bias), where experience had revealed that standards were often appalling and convenience took precedence over welfare.

Products and training included the introduction of ethics modules, guidelines and codes of practice, and the development of partnerships with DfEE and LEAs.

With the introduction of the National Curriculum in the late 1980s and science becoming a core subject, there were demands for a more clearly focused science approach and a move to identify broader ecological issues than had been addressed in the 1970s. Primary education received an ever-increasing focus, because RSPCA materials lent themselves to the cross-curricular approach that many schools retained despite the outwardly fragmented nature of the National Curriculum. There was a recognised need on the part of non-specialist primary teachers to receive science-specific in-service training (Inset) and the organisation was receiving an increased number of requests for support from primary schools. By 1995, the RSPCA had produced a National Curriculum links card, referencing RSPCA-based activity to the National Curriculum Programmes of Study, and a post-Dearing update with extended reference to citizenship, very much in vogue at the time. The year 1996 saw the publication of the *Animals and Education Policy Planner*, which was initially aimed at school governors and, later, head teachers. This supported schools in producing a school policy which would then have RSPCA accreditation. This was followed in 1998 by the *Good Practice Guide*, which reflected the outcome of evaluation and celebrated good practice in schools. Its less 'evangelical' format (i.e. 'this is what schools are doing', not 'this is what we are telling you to do') was considered more accessible.

It is interesting to note that there has been an association between the ASE and RSPCA for many years, and, while their views have at times conflicted, this has proved a useful vehicle for promoting both welfare issues and the broader picture in the context of science. Initial teacher training (ITT) is another area that was seen as an area ripe for development but, as with other areas of education, the perceived time-restrictive nature of the National Curriculum limited the amount of work that was done.

Industrial Corporations: Glaxo Wellcome

So far I have considered the work of charitable or sponsored organisations whose work is perceived as being broadly altruistic. Using particular examples, I shall now turn to the role that industrial and commercial organisations play in supporting and influencing the progress of science education in primary schools. The first consideration in this area is the place of the individual corporation, using as the example Glaxo Wellcome. Their stated rationale with regards to primary science education is to:

- encourage interest in science from an early age;
- improve public understanding of science and the image of industry;
- support teachers in achieving these aims;
- improve the quality of the work force in terms of both the general perceptions of science and specialist scientific skills.

In fulfilling these aims, there is agreement that there should be no product or operation promotion in primary schools.

The principal focus began at the end of the 1980s, with a 'downward push' in terms of the children being supported from existing secondary initiatives. The timing was coincidental to the introduction of the National Curriculum but this inevitably provided an increased focus. The later development of the Channel 4 television series *Making Sense of Science*, with materials and training supported by Glaxo Wellcome, both provides an exemplar of how industry can work with others to support science in school and also illustrates the increasingly 'primary friendly' nature of supported initiatives. In developing the series, the Department for Trade and Industry (DTI) had initiated development of a programme which reflected many of the constructivist principles that were evident in the SPACE research. The programme was produced to support teachers of the 7–11 age range. Programmes addressed teaching methodology, underlying subject knowledge and understanding, and children's misconceptions, and gave examples of appropriate contexts within which to teach specific aspects of science. Recognising the need to support the programmes with training, the DTI looked for a major sponsor for the second phase, in-service provision. Glaxo Wellcome took up the challenge of supporting the development of a video/teachers' pack and in-service package. There was a near 100 per cent uptake among those schools offered training opportunities. Teachers recognised the quality of the product and the training. Both the content and the proposed methodology were perceived as relevant. DfEE perceptions of the constituents of 'good practice' – such as effective teacher questioning – were woven into the materials, and this combination of relevant content and sound pedagogy was seen by teachers as extremely valuable. Core training events involved 3,000 co-ordinators (cascading to staff in their schools), and SATRO- and LEA-run events expanded this number by a further 2,000.

Involvement with the *Making Sense of Science* project also gave the education officers opportunities to work with other individuals involved in promoting primary science through ITT and in-service training. For example, the concept cartoon project *Science on the Underground*, which Glaxo acknowledge as one of their most successful public understanding of science initiatives, came to their notice while working with ITT lecturers from Manchester Metropolitan University, who had already developed the cartoon materials under the heading 'Starting Points for Science'.

It will have been evident that industrial organisations such as Glaxo Wellcome do not operate in isolation, and the majority of industry-based sponsors turn to liaison agencies to inform and disseminate. I shall look at two such organisations. The first, Science and Technology Regional Organisations (SATRO), stemmed from government initiatives but now operate as independent agencies, while the second, the Chemical Industry Education Centre (CIEC), is directly supported by the chemical industry.

Industrial Liaison Organisations 1: Science and Technology Regional Organisations

First steps in the development of SATROs were in the late 1950s, and were made in the education departments of HEI institutions by like-minded colleagues who recognised the potential benefits of working with industry. The main focus at the time was on secondary education. However, the primary sector has received ever-increasing attention since the 1980s. The co-ordination and establishment of a formal network of SATROs came through the initiative of the Standing Conference for Schools' Science and Technology (SCSST), founded in the 1960s. This was a jointly funded venture between the DTI and industry, both of whom were anxious about the direction in which science and technology were going in terms of research, production and, in particular, recruitment. Today, of the forty-three established SATROs, a few remain linked to HEI institutions. A similar proportion are linked to LEAs. A third and growing number are called Education and Business Partnerships (EBP) and are funded from the local Training and Enterprise Councils (TEC). The remainder are independent limited companies whose entire funding comes from external sponsorship or contract bidding.

What is the motivation behind these initiatives? On the largest scale, it is a desire on the part of industry to improve the quality of the work force – an encouragingly long-term vision with minor short-term pay-back. The other motivation is to raise the perspective of industry among the young, generating a positive view of what industry has to offer in terms of both product and employment, and to flag up the role industry plays in community relations. The principal means of achieving these goals is through the curriculum and through training.

SATROs understand both the economic needs of the locality and the regional development organisations' national initiatives. They also recognise what schools need, based on government education agenda and Ofsted reports, and act as brokers/facilitators to make things happen. The philosophy is clear; SATROs aim:

- to influence young people (4 to 19);
- to ensure the majority of educators and their pupils are aware of what industry has to offer;
- to encourage stronger vocational elements to science and technology education with a commitment to 'active learning'.

In fulfilling these aims, SATROs hope to provide enhancement and enrichment for children's learning science and technology, and to provide enrichment for teachers through classroom support, updating Inset and professional development. In terms of influence at a national policy level, the National Curriculum Council (NCC), the Qualifications and Curriculum Authority (QCA) and the Teacher Training Agency (TTA) have all consulted with what was SCSST, now called the Association of Schools' Science, Engineering and Technology (ASSET). In fact, a seconded representative of the DfEE is on the staff of ASSET, thereby ensuring dialogue between SATROs and government.

Industrial Liaison Organisations 2: Chemical Industry Education Centre (CIEC)

The CIEC, based in the Department of Chemistry at the University of York, was established in order to improve mutual understanding between schools and the chemical industry so that teachers and industrialists would have a clearer insight into each other's needs. It operates by providing schools with a range of resources that are clearly targeted at the National Curricula of England, Wales and Scotland. These materials are all developed by practising teachers working alongside industrialists. One important characteristic is the balanced coverage that they offer between industrial and environmental issues. They reflect a positive response to critical comments, such as those made during a survey of teachers' opinions carried out for this chapter and discussed later. In particular, there is clear consideration of the need to offer differentiated materials which are appropriate for the age being addressed. The most recent innovation is the Primary Partners project, which aims to facilitate partnership between primary schools and industry to enhance science teaching. The supporting pack of materials clearly reflects the Chemical Education Group's thinking. This is expressed in the Salters seminar report of November 1996, in which the following statement was identified as a key issue with respect to primary science: 'There is a need to support the teaching of science in primary schools and, to this end, promote links between industry and primary schools' (Chemical Education Group, 1996).

Industrial Societies: The Society of Chemical Industry (SCI)

SCI sees the development and direction of science teaching and learning as fundamental to the development and direction of science in industry. SCI is a global interdisciplinary network with deep roots in business, manufacturing, consumer affairs, research and education at all levels. Among SCI's many aims is to make science more accessible and stimulating for both pupils and teachers. SCI prides itself on a long heritage of strong industrial links, but would perhaps be better described as a charity with scientific and industrial interests than an interest group with industrial links. Its significant contribution to primary science education was the establishment of SCIcentre, a collaborative project between the University of Leicester School of Education and Homerton College, Cambridge, initiated and funded by the Society of Chemical Industry. SCIcentre is SCI's millennium project. It aims to encourage mutually beneficial collaboration between organisations involved with initial teacher training to increase output of good, newly qualified primary school teachers with regards to science.

In addition to establishing SCIcentre, SCI has provided a range of other opportunities to support primary science, including the production of Christmas lectures specifically aimed at primary schools, participation in the BBC film series *The Science of Almost Everything*, and provision of the Schools Material Packs – resource packs available on request for primary schools. So far, it is clear that the nature of support for and influence on primary science education is diverse and extensive. Yet despite the diversity there are common themes; the move towards a scientifically literate society, the public advancement of science and the provision of useful and practical support for teachers.

Future Initiatives

I have attempted to establish the background that underlies the intentions and operations of non-governmental interest groups with regard to the development and support of science over the last two decades. What, then, is their future role?

The Nuffield Foundation makes some clear recommendations in *Beyond 2000, Science Education for the Future* (Millar and Osborne, 1998). The report was the synthesis of a series of funded seminars drawing on the opinions and expertise of leaders in science education. It sprang from the desire to

> provide a new vision of an education in science for our young people and driven by a sense of growing disparity between the science education provided in our schools and the needs and interests of the children who will be our future citizens.
>
> (Millar and Osborne, 1998, p. 1)

The notion of a substantive re-think regarding the direction that science education should take came, paradoxically, during a period of government-enforced stability – the 1994 moratorium on curriculum change to the year 2000. However, the Nuffield Foundation felt it far better to be visionary rather than providing the reactive response to 'top down' curriculum implementation of the previous ten years. It was felt that change in the future should have its foundation in the kind of reflective, research-informed debate that characterises the report. This is not to say that what the report recommends is necessarily 'the word'. Even on the day of publication, serious misgivings regarding the direction that some of the proposals indicated were voiced by both industrialists and educationalists present. It was clear that any influence the Nuffield Foundation hoped to wield over future government policy would result from the outcome of the debate the report promoted rather than the contents of the report itself. The overarching issue for the seminars that informed the report was the dichotomy between the aim of science educators to provide an adequate science 'training' for a minority of future scientists and the aim to develop a scientifically literate majority.

There is little doubt that as we move into the future we are likely to see even greater advances in science. The idea of a society that is scientifically illiterate becomes one of increasing concern – 'positively dangerous', as one delegate to the Nuffield seminars that fed into *Beyond 2000* put it. There is a tension in scientific progress between the beneficial outcomes of science and the unintended consequences and ethical issues that ensue. As this Faustian bargaining comes more into the public arena, so the pressure on science education to produce a more clearly informed and interested society increases. Science education must teach society what science is really about: not necessarily knowing, but being able to think clearly, and framing the right questions to move closer towards understanding. There is a genuine concern that associated with poor scientific literacy is a failure to make informed personal decisions related to health and safety and, on a larger scale, to engage with wider issues in science such as genetic modification of foodstuffs. While recognising how far things had come over the previous twenty years, particularly in allowing access to scientific ideas and processes for the very young, the report identified 'remaining problems'. It felt that many young people complete compulsory science education with apparent success, yet lack familiarity with the scientific ideas likely to be encountered outside school. School science, particularly at the secondary level, often failed to engender the natural curiosity and inquisitiveness characteristic of many primary school children. The lack of relevance of the 1995 National Curriculum to the 'real' issues regarding science was identified as the key element, the 'turn-off' factor. This lack of relevance stemmed from a curriculum which lacked clear aims in terms of scientific literacy or an agreed model of pupil development over the 5–16 age range. The curriculum could be characterised as being comprised of discrete ideas, with assessment based largely on recall rather than on application of concepts and procedures. There was little emphasis on discussion or analysis of any of the scientific ideas that permeate contemporary life. In many respects it was a National Curriculum that separated science and technology, despite most people's perception that scientific endeavour is reflected in technological product. Even the innovative aspect of encouraging investigations in Sc1 of the National Curriculum was perceived as in danger of succumbing to over-systematised teaching to meet specific assessment criteria.

To address these problems, the report made ten recommendations. Those relevant to primary and middle schools are:

- The science curriculum from 5 to 16 should be seen primarily as a course to enhance general scientific literacy.
- The science curriculum needs to contain a clear statement of its aims – making clear why we consider it valuable for all our young people to study science and what we would wish them to gain from the experience. These aims need to be clear, and easily understood by teachers, pupils and parents. They also need to be realistic and achievable.
- The curriculum needs to be presented clearly and simply, and its contents need to be seen to follow from the statement of aims. Scientific knowledge can best be presented in the curriculum as a number of key [representative] 'explanatory stories'.
- Work should be undertaken to explore how aspects of technology and the applications of science currently omitted could be incorporated within a science curriculum designed to enhance scientific literacy.
- The science curriculum should provide young people with an understanding of some key [representative] 'ideas-about-science', that is, ideas about the ways in which reliable knowledge of the natural world has been, and is being, obtained.
- The science curriculum should encourage the use of a wide variety of teaching methods and approaches. There should be variation in the pace at which ideas are introduced. In particular, case studies of historical and current issues should be used to consolidate understanding of the 'explanatory stories' and of key ideas-about-science, and to make it easier for teachers to match work to the needs and interests of learners.
- The assessment approaches used to report on pupils' performance should encourage teachers to focus on pupils' ability to understand and interpret scientific information, and to discuss controversial issues, as well as on their knowledge and understanding of scientific ideas.

(based on Millar and Osborne, 1998)

In the short term, it was suggested that those aspects of the general requirements of the National Curriculum (Sc0) should be incorporated into the first Attainment Target (Sc1) to give more stress to the teaching of 'ideas-about-science'. In addition, new forms of assessment should be developed to reflect such an emphasis. In the medium and long term, the recommendations were that a formal procedure should be established whereby innovative approaches in science education are trailed on a restricted scale in a representative range of schools for a fixed period. Such innovations should then be evaluated and the outcomes used to inform subsequent changes at a national level. Certainly, no significant changes should be made to the National Curriculum or its assessment unless they have been piloted in this way.

The qualification of the final recommendation regarding the evaluation of a research-based approach to curriculum development, and the necessity to pilot fully such innovations in approach before either advising or legislating change within the teaching profession as a whole, must be considered welcome. It recognises the problems that dogged both the implementation of the original National Curriculum and the extrapolation of research findings into practice. The new statutory orders for science do go some small way to meet the recommendations of the *Beyond 2000* report. Equally, it is apparent that it is teachers who inevitably will do most to meet the report's recommendations, rather than the new curriculum.

How does the Nuffield Foundation's view fit with the other non-governmental, non-commercial or 'single-issue' interest group under scrutiny? One answer to this question comes in the ASE 'mission statement' (ASE, 1998). The consistent message from the consultation that informed the document was the importance of the place of science in our lives, the recognition of the value of scientific literacy as an essential requirement of citizenship in the twenty-first century. The report identifies the need to explore appropriate and creative assessment methods and to reduce/re-examine the content of the curriculum as key factors in shifting the balance of science education towards a notion of 'scientific literacy'.

The notion they use of developing key ideas through a 'story-line' approach relates closely with the 'explanatory stories of science' idea developed in the Nuffield Foundation recommendations. The key ideas should enable learners to make sense of science, and have some kind of global significance rather than simply being a list of facts. Equally, the ideas should have personal significance and be supported by practical activities which reinforce the ideas and allow their application to new and relevant situations. Perhaps the most crucial element of the recommendations is the notion of mapping progression in a core of knowledge and understanding which enables children to gain a sense of how ideas develop and to recognise the importance of their own ideas in their learning. This meta-cognitive element of the constructivist approach is further explored in Chapter 4.

Reflecting on the role of the other groups under consideration, how do these organisations view their role in the future development of primary science? The evidence is that they will continue to support whatever innovations are made at a government level, with appropriate support materials for teachers based on the advice of internal education officers or external consultancy from ITT or LEA advisory services. Characteristically, the overarching nature of what these organisations have offered teachers and learners to date has been by way of curriculum enrichment, providing a 'real life' context and making some attempt to fill the gaps between the facts that prevent the current curriculum from being perceived as holistic. Often, this 'science in the real world' tone that is particularly characteristic of industry's attempts to support and supplement primary science can be taken as a tacit critique of the perceived 'science in a vacuum' experience that children receive at school. It will be interesting to see the reaction of industry should the recommendations of the ASE and Nuffield be taken on board by the QCA and government. What would be the implications of a curriculum that focused more on 'scientific literacy' than on 'content'?

In the research for this chapter, industry and the QCA have criticised the contextual desert in which much science in the National Curriculum takes place. In many classrooms, only lip service was paid to the general requirements (Sc0) of the 1995 version of Science in the National Curriculum. Yet this could have provided a foundation for the holistic approaches that the ASE and Nuffield seminar group support. Sc0 was not assessed and therefore was not perceived by many teachers as a priority.

Recognising the Needs of Teachers

Above all, recognition must be made of teachers' needs and perceptions. In the survey carried out for this chapter of over one hundred schools in East Anglia, an area well served by SATRO, local industry and an active regional ASE group, three somewhat alarming patterns arose.

It was clear that many teachers were simply unaware of non-governmental agency support and initiatives or, if they were, only of specific 'products'. It was equally evident that only a minority consulted the range of available organisations as a matter of course when planning. This is of added concern in the light of the research which indicated that those who did avail themselves of support were very happy with what was offered. In other words, the poor uptake was not an issue of quality but of communication.

The second anxiety, which almost 90 per cent of respondents referred to, was the failure of materials produced by non-governmental agencies to fit with the complex and somewhat inflexible whole-school planning characteristic of the late 1990s. At best, they might offer some 'research' extension work for individual children's topics, but essentially there appears to be no room for the current range of support materials.

The final area of concern was linked to the target audience for materials. Most had a clear Key Stage 2 focus, and those that purported to be aimed at lower primary were, in fact, inappropriate. This issue was best expressed by a Cambridgeshire nursery head teacher who wrote:

> I wish I could feel that the educational advisers to these packs [referring to industry-sponsored educational support materials] still had hands-on classroom experience in the Early Years rather than 'watering down' work for older children and thinking that is appropriate.

This is a criticism from many representatives of the early years – local advisers, heads, education officers and teachers. To be fair, all the organisations that were approached for comments for this chapter understand the growing need to provide support for Key Stage 1 and the early years.

The other area that all the organisations identify for the future, which will hopefully address the first of the concerns, is that of initial teacher training. While inevitably constrained by the same 'curriculum overload' that restricts the way support is taken up in schools, it is important that trainee teachers are made aware of how the non-governmental agencies operate and are encouraged to consider what is offered as a matter of course when planning.

References

Association for Science Education (1998) *Science Education for the Year 2000 and Beyond*, Hatfield: ASE.

British Association (1998) *Facing the Future*, London: BAAS.

Central Advisory Council for Education [England] (1967) *Children and their Primary Schools* (the Plowden Report), London: HMSO.

Chemical Education Group (1996) *Primary Science Seminar Report*, London: Salters.

Chemical Industry Education Centre (1996) *Support for Science Education*, York: CIEC.

Chemical Industry Education Centre (1999) *Primary Partners*, York: CIEC.

Dearing, R. (1993) *The National Curriculum and its Assessment: Final Report*, London: SCAA.

DfE (1995) *Science in the National Curriculum*, London: HMSO.

Harlen, W. and Darwin, A. (1977) *Match and Mismatch*, Edinburgh: Oliver and Boyd.

Keogh, B. and Naylor, S. (1997a) *Starting Points for Science*, Sandbach: Millgate House.

Keogh, B. and Naylor, S. (1997b) *Thinking About Science Concept Cartoons*, Sandbach: Millgate House.

Mid Anglia SATRO (1998) *Business Plan 1998/99*, Cambridge: SATRO.

Millar, R. and Osborne, J. (1998) *Beyond 2000, Science Education for the Future*, London: Nuffield Foundation.

Nuffield Foundation (1997) *Annual Report*, London: Nuffield Foundation.

Parkin, T. and Lewis, M. (1998) *Nuffield Exeter Extending Literacy (EXEL) Project. Science and Literacy*, London: Collins Educational.

Richards, R., Ennever, L., James, A. and Harlen, W. (1972) *Early Experiences: A Unit for Teachers*, London: Macdonald Education.

RSPCA (1995) *National Curriculum Links*, Horsham: RSPCA.

RSPCA (1997a) *Good Practice in Animal Welfare Education*, Horsham: RSPCA.

RSPCA (1997b) *Animal Care: Early Years Resource Pack*, Horsham: RSPCA.

SPACE Research Reports (various dates) e.g. Russell, T. and Watt, D. (1990) *SPACE Research Reports: Growth*, Liverpool: Liverpool University Press.

Stringer, J. (1995) *Making Sense of Science: Teachers' Materials*, London: Channel 4 Schools.

10 Science in Society or Society in Science?

Michael Reiss

This chapter begins with a brief history of science-in-society type courses in UK science education. While such courses were initially confined to secondary schools and colleges, they subsequently reached primary and middle schools. However, these courses have not had the impact that might have been anticipated. I examine the reasons for this. I then argue that a richer understanding of the nature of science than often obtained leads to a science education more suited for all pupils. In particular, I explore gender issues, multicultural and anti-racist science education, and science education for pupils with disabilities.

The Arrival of Science-in-Society Courses

The history of UK school science curricula has largely been one of small groups of committed science educators launching new teaching programmes. These have mostly been produced because of a single, critical perceived weakness in existing programmes. They are typically backed up by written materials, claims about a novel approach to the teaching of science and, for some age groups, a shift in the form(s) of assessment.

Thus, in the 1970s and 1980s, concerns about the perceived failure of existing school science lessons to tackle applied aspects of science adequately or to relate the science taught in school to the wider needs of citizens in society led to a number of new science programmes aimed at the secondary student. Notable among these were the Science in Society programme, the SISCON (Science in a Social Context) programme and the various SATIS (Science and Technology) courses.

The Science in Society programme was launched in 1981 (Association for Science Education, 1981). It consisted of twelve readers, each about fifty A5 pages in length, and a teacher's guide. Each reader contained a number of short articles contributed by prominent figures in industry, the professions, the academic world and in politics. Most significantly, as stated on the back of each reader, the aim was to:

> provide a new course in science and society for general studies at sixth-form level. The course has been specially designed to make scientific problems accessible to the non-scientist, as well as to explain the social aspects of science to the scientist.

Twenty years later, the Science in Society programme inevitably looks somewhat dated. The first reader in the series, *Diseases and the Doctor*, has, for example, a four-page chapter on 'Diseases of women', sandwiched between a chapter on 'Epilepsy' and one on 'Drug and alcohol dependence'. The 'Diseases of women' contribution is just the thing to enthuse 16–19-year-olds. The section titled 'The adolescent girl' is entirely devoted to 'problems', namely late onset of menstruation, dysmenorrhoea, irregular/heavy menstruation and amenorrhoea.

Hot on the heels of the Science in Society programme was SISCON (Solomon, 1983). Again intended for general studies at sixth-form level, though with the somewhat surprising provision that it could also be examined as a Mode 3 CSE, this consisted of eight readers, each about forty A5 pages in length, and a teacher's guide. SISCON had virtually identical aims to the Science in Society programme.

In addition to any good they did in themselves, the Science in Society and SISCON programmes helped spread the notion within the science education community that materials were needed to help pupils appreciate the interactions between science and society. In particular, the Association for Science Education gave its blessing to SATIS. This successful and long-running project began with materials for 16–19-year-olds but soon spread to 14–16-year-olds and then to 8–14-year-olds (Association for Science Education, 1992–3).

The Effect on the Primary Curriculum

The SATIS 8–14 Project consists of three boxes of materials. Box 2, for example, contains ten books, each about fifty A4 pages in length, a teacher book and a computer disk. All the materials are designed to ensure relevance for the English and Welsh Science and Technology National Curricula as well as curriculum recommendations for Northern Ireland and Scotland. For example, a unit titled 'Lettuces for Profit' looks at ways of controlling crop growth to produce plants at the right time and to maximise yield; a unit titled 'Does Your Town Need a Bypass?' gets pupils to consider the problems posed by holiday traffic through an imaginary village called Piggleswick; and a unit titled 'Fireworks' includes everything from flame tests and a history of fireworks to the analysis of firework injury data and a major role play about whether families in Bridgehouse – an imaginary town in the north of England – should have their own home-based firework parties or go to the large display put on by the local Asian community to celebrate Diwali.

Such initiatives, together with the requirements of the Science National Curriculum (in its various incarnations since 1989) for pupils to explore social aspects of science, might have caused one to hope that by now the various primary and middle school teaching materials on offer to teachers would provide a rich picture of science. However, the early opportunities afforded by Attainment Target 17, titled 'The Nature of Science', were somewhat lost as subsequent revisions of the Science National Curriculum presented a more conventional picture of science. Nevertheless, the new statutory orders for science require, at KS1, pupils to be 'taught skills, knowledge and understanding through the use of a range of domestic and environmental contexts that are familiar and of interest to them' (QCA, 1999), while at KS2 the equivalent injunction states that pupils should, in addition, be taught through 'everyday' and 'local' contexts.

Yet few are the primary science classrooms where pupils spend a high proportion of their science lessons engaged on SATIS-like activities. For all the apparent encouragement of the above Science National Curriculum statements, why is so much school science so different?

There are several answers to this question. For a start, while the topic approach to much primary education in the 1960s, 1970s and 1980s often provided relevant contexts for children to undertake their work in science, the sheer volume of material that had to be covered once the National Curriculum was introduced led many teachers to jettison these everyday, meaningful contexts for their science teaching.

A second reason is that the National Curriculum science tests (or SATs, as they are still generally known) at the end of KS2 mostly present a 'pure' vision of science, a science largely divorced from technology and the everyday concerns of people, a science that occupies itself not so much with profitable lettuces, town bypasses and firework safety (à la SATIS) but with chains of paper clips (forces), Janet-and-John-like gardens (food chains) and the endless excitement of droplets of water condensing on cold glass (changing materials).

As someone who sat for several years on the vetting panel that helped set these tests, I acknowledge the difficulties of trying to devise more meaningful questions. Any attempt to introduce a note of reality into a question almost invariably led to howls of protests either along the lines of 'Teachers won't be expecting that' or on the theme of 'We can't produce an unambiguous answer for that.' In such circumstances, question-setters and schools become locked in a framework from which attempts to escape to the rest of the universe are unlikely to succeed.

A third answer to the question is that most science teachers have a view of science which sees 'true' science as the pure aspects of the subject. Working out how to grow lettuces for profit is assumed to be something that should be done *after* one has waded through the minutiae of plant structure and growth. The problem, of course, is that today's primary curriculum simply doesn't allow time for such luxuries. No sooner has a teacher finished plant structure and growth than she is on to the next area of the Programme of Study, with all the joys contained therein. Indeed, there is some evidence to suggest that, in primary schools, references to out-of-school experiences are usually initiated by pupils rather than teachers (Edwards and Mercer, 1987). In secondary science classrooms, links to out-of-school experiences are more likely to be made by teachers, but in any case are made only rarely (Mayoh and Knutton, 1997).

The Relationship between 'Science' and 'Society'

Science-in-society courses have much to commend them, though they have sometimes laid themselves open to accusations of 'dumbing down' science. At their best, they provide a more meaningful science education and one that, by being less abstract, is likely to engage the interests of a wider range of those tackling science courses as pupils and students (see Solomon, 1993).

But these science-in-society courses have another consequence. They help to raise deeper questions about the connections between science and society – hence the title of this chapter. It must clearly have been the case that some of those teaching such courses initially assumed, no doubt implicitly, that the relationship between science and society was a simple one. Nevertheless, the very juxtaposition of the terms 'science' and 'society' causes one to examine their relationship more exactly.

A standard understanding of how science advances, such as might be held by a 'person in the street' if this person has such a conception, probably runs something like this:

> Science is carried out by scientists. Scientists investigate questions about the world around us. They come up with ideas and then test them.

Such a view of science is likely to be reinforced by anyone surviving eleven years of the four Attainment Targets of the Science National Curriculum in England and Wales. A major research project titled 'The Development of Pupils' Understanding of the Nature of Science' and carried out between 1991 and 1993 on 9-, 12- and 16-year-olds came up with the following findings (Driver et al., 1996):

- Students tend to see the purpose of science as providing solutions to technical problems rather than providing more powerful explanations.
- Students rarely appreciate that scientific explanations can involve postulating models. Even when they do, models are presumed to map on to events in the world in an unproblematic manner.
- Students rarely see science as a social enterprise. Scientists are seen as individuals working in isolation.
- Students have little awareness of the ways that society influences decisions about research agendas. The most common view is that scientists, through their personal altruism, choose to work in particular problems of concern to society.

There is little in either the 1991 or the 1995 version of Sc1 in the Science National Curriculum or, indeed, within the new statutory orders for science, to encourage one to expect that students are likely to acquire a more nuanced view of what science is and how it is done (cf. Donnelly et al., 1996).

The Nature of Science

For many years the fundamental question about the UK science curriculum was whether it should focus on the findings of science (its content) or on how science reaches its conclusions (its process). Although academic historians and philosophers of science and science educators generally accept that this dichotomy is, at best, overstated and possibly meaningless, these two approaches sit in apparent isolation in the Science National Curriculum as Sc1 (process) versus Sc2, Sc3 and Sc4 (content).

Not only that, but Sc1, for all its encouraging emphasis on investigative activity, paints a very narrow picture of what it is to do science. Assessing the work of Darwin in his *The Origin of Species by Means of Natural Selection* and Mendel's life work by the criteria of AT1 of the 1991 Science National Curriculum gives Mendel a Level 5 and Darwin a Level 3 (Reiss, 1993a). Mendel's problem was that he chose only to investigate variables that varied discontinuously, whereas Level 6 required continuous variables. Poor old Darwin never even used a 'fair test' in his pioneering work, so he fails to reach Level 4.

Actually, carrying out a fair test is almost the only thing that most pupil investigations at KS2 and KS3 consist of (Goldsworthy, 1998). When I was at school I remember a teacher once asking us what distinguished poetry from prose. After a welter of possible answers from his pupils he informed us, in all seriousness, that the answer was that in poetry each line begins with a capital letter. (It was a very expensive fee-paying scho ol.) Perhaps it is time for pupils to realise that, just as there isn't one way of writing poetry, so there isn't one way of doing science. It is perfectly understandable why many primary teachers, particularly if they are not science specialists, may present pupils with a somewhat formulaic approach to 'doing science'. However, it needs to be stressed that there is no such thing as 'the' scientific method. This is not to equate science with poetry or with mathematics or basket-making or with any other area of human knowledge and expertise. Rather, it is to acknowledge what both the lesson of history and the current diversity among scientists teach us, namely that there are a variety of scientific methods.

For a balanced defence of this position, see almost any significant current work in the history and philosophy of science (e.g. Chalmers, 1990; Cunningham and Williams, 1993). For a distinctly off-beam but far more entertaining articulation of this position, here are some quotes from Paul Feyerabend, who was Professor of Philosophy in the University of California at Berkeley and Professor of the Philosophy of Science at the Federal Institute of Technology at Zürich:

> the events, procedures and results that constitute the sciences have no common structure.
>
> (Feyerabend, 1988, p. 1)

> the success of 'science' cannot be used as an argument for treating as yet unsolved problems in a standardised way.
>
> (ibid., p. 2)

> there can be many different kinds of science. People starting from different social backgrounds will approach the world in different ways and learn different things about it.
>
> (ibid., p. 3)

Most scientists have never heard of Feyerabend and would rubbish his arguments if they had. The one philosopher of science most scientists like is Karl Popper. This is because Popper saw the key to science as being the principle of falsification. For Popper, a meaningful scientific statement is one that can be falsified. If a statement or theory is not capable of being falsified, it is not scientific.

While there is much of value in Popper, subsequent research has shown that although scientists may like to think they are Popperians, they are not. Scientists are actually human. They do not like to abandon their pet beliefs any more than the rest of us do, and, in particular, they do not come to disbelieve a theory simply because there is evidence against it. Rather they treat theories like stepping stones, only abandoning one when there is another better one on to which they can hop (see Kuhn, 1970). This is true of pupils, too.

Science Education for All

Gender Issues

The last twenty years have seen a profusion of scholarship about gender issues in science education. At first, such writings tried to address the perceived problem of not enough women going into science, especially the physical sciences, by suggesting that science curricula contain more girl-friendly topics, such as growing crystals (in chemistry) and the electronics of domestic machines (in physics). With the benefit of hindsight, it is now thought that such approaches, and many of the others undertaken in the 1980s, focused too narrowly on attempts to alter the option and career choices made by young women (Byrne, 1993; Henwood, 1996).

More recently, research has shifted to the nature of pedagogy and to the effects of girls and boys on each other's learning and attitudes towards certain subjects. Considerable interest has been aroused by innovations where some subjects, such as science, mathematics and languages, have been taught in single-sex groupings even within mixed-sex schools. The results are as yet inconclusive, though encouraging (Kruse, 1996) but the mere fact that increasing numbers of schools are undertaking such experiments, in some cases encouraged by Ofsted training materials, is encouraging in that it bears witness to the steps schools are prepared to take to reduce inequalities.

It is increasingly recognised how science texts may present different, biased messages about what it is to be female and what it is to be male, and how teachers may differentially reward and react to students depending on their gender (Guzzetti and Williams, 1996). While much of this work has simply emphasised the extent to which female pupils are (whether consciously or not) discriminated against, psychologists have explored the extent to which males and females have different ways of knowing.

Such work is controversial, for one important strand of feminism has been to argue that there are no essential differences between men and women, only differences that result from our having been brought up in a patriarchal society. Nevertheless, the classic work by Gilligan (1982) has made it easier for educationalists to explore what might be the consequences of there being sex-specific differences in cognitive style.

Head (1996) provides a particularly clear overview. He argues that there are four safe generalisations that can on average be made:

- Females tend to embed information in its context; males tend to extract it from its context.
- Females are more reflective, males more impulsive.
- When something goes wrong, females are more likely to blame themselves, males to locate the blame elsewhere.
- Females are more likely to co-operate, males to compete.

Now, the consequences of this for the teaching of science in primary and middle schools are several. First, it needs to be stressed that there are many exceptions to such generalisations. There are girls who have a penchant for analytical rather than synthetic thinking; there are boys who blame themselves. Nevertheless, any experienced teacher is likely to recognise the considerable truth in such generalisations. The more interesting question is: what should we do with them?

Take, for example, the first of Head's points – that females tend to embed information in its context while males tend to extract it from its context. Head argues that there are times in science when one needs to extract information from its context and there are times when one needs to embed it in its context (an exemplification of Feyerabend's point above that 'there can be many different kinds of science'). In other words, both extracting and embedding are skills that are of value in science. It may well be that an individual pupil naturally gravitates to one of these as a preferred style, but each is of value and needs to be taught.

Similarly, a successful teacher at times encourages co-operation (indeed, given the perennial shortages of science equipment in schools, just about all practical work in science is 'co-operative') and at times allows or encourages a certain amount of competition.

Multicultural and Anti-racist Science Education

There are still those who argue that true science arose only once, around the seventeenth century, and in one place, Western Europe. It is true that such voices come from those who have read little history and philosophy of science, being instead scientists. Nevertheless, the general public is more likely to buy the books of scientists than of historians, philosophers or sociologists of science, and the 'one, true, recent science' argument, as advocated, for example by Wolpert (1992) provides an easily understood message.

From an educational point of view, I believe it is important to argue for the following two positions. First, the antecedents of modern science are found in every culture. All people have a need to understand, to explain and to control the workings of the natural world. All cultures show both scientific thinking and technological expertise. Second, today's science is not the preserve of the West. Modern science is carried out by peoples of every nationality, creed, ethnicity and culture.

Indeed, I have argued that one helpful approach is to see science as a collection of ethnosciences (Reiss, 1993b). This makes it easier to reject the 'one, true, recent science' argument. Once science is seen as consisting of a number of sciences, the focus of one's science teaching shifts somewhat (see Peacock, 1991; Reiss, 1993b; Byers et al., 1994; Siraj-Blatchford, 1996; and Reiss, 1998).

An analogy with music teaching may help. Broadly speaking, a music teacher has two aims. She hopes that pupils will derive a fuller appreciation of the various forms of good music, and she hopes that they will each develop their ability to compose music, to sing and to play a range of instruments. Note that she does not (hopefully) believe that there is only a single good form of music – as if all music other than classical was bad – and note that she (hopefully) welcomes a diversity of pupils in her class, since a diversity of pupils makes it easier for pupils to learn from each other about the diversity of musics that there are.

Science teaching can be the same. There is no one perfect way of separating substances or classifying organisms or devising a circuit so that a buzzer sounds when it gets cold. Of course, some ways of doing these things are better than others, just as some ways of playing the recorder are better than others. Similarly, there aren't many ways of labelling the parts of a buttercup, just as there aren't many ways of getting a piano to play middle C.

Does the argument I have been advancing against the idea that there is one true science mean that a teacher should never steer a child in a particular direction in a science lesson? Of course not. That would be to abdicate one's role as a teacher. Realising that science is a collection of ethnosciences does not mean a total abandonment of all objectivity and a necessary descent into some anarchic post-modern nihilism. There is good science and there is bad science, just as there is good painting and bad painting. But as there are many genres of paintings, so there are many methods of science.

Science Education for Pupils with Disabilities

There are two main categories of mistake that can be made when educating pupils with disabilities, whether mental, physical, behavioural or emotional. One is to assume that such pupils cannot cope with normal teaching; the other is to assume that such pupils need only the usual teaching. An extreme illustration of the former tendency is provided by the story of Anne McDonald, who has cerebral palsy. As a youngster, doctors had judged her to have a mental age of 1. She was force-fed and spent her days lying on the floor. As she later wrote:

> I went to St Nicholas Hospital when I was three. The hospital was the state garbage bin where very young children were taken into permanent care …
>
> Seeing the occasional television programme gave me some ideas. The Bronowski *Ascent of Man* programmes were critical in bringing me in contact with scientific method … When Joey [another child labelled 'profoundly retarded'] taught us about fractions, suddenly everything started to come together. I started doing arithmetic for fun. I also tried to work out some constants. I had a go at the speed of light, using the distance of the moon from the earth (which had been given coverage during the Apollo missions) …
>
> Bronowski covered Pythagoras and I had ample opportunity to think about the implications. The hospital nappies were not square, and every time the nurse had to fold a nappy they had to square it first. I became aware of symmetry and its importance in geometry. To calculate I used a crude abacus based on the clock. I used to work in base twelve.…
>
> Sometimes I was hit because I talked with other children, and the nurses thought I was screaming without reason. Since we were always with nurses opportunities for speech were few.
>
> (Baird, 1992, p. 6)

The ASE's policy on 'Access to science education' is helpful:

> Learners can be isolated by sensory, physical, cognitive and emotional problems. Choosing familiar contexts and providing appropriate activities motivates and stimulates learners.
>
> Appropriate science experiences will involve:
>
> - using a range of teaching and learning strategies
> - developing concepts and skills gradually
> - matching the demands of the activity to the learner
> - allowing different outcomes for different individuals
> - building on the learner's strengths
> - allowing time for learners to reflect on their work
> - using a range of methods to monitor progress
> - ensuring written material is at an appropriate level for each learner
> - explaining new vocabulary

- using first hand examples to reinforce understanding
- using a range of communication methods
- adopting a consistent presentation style for written work
- ensuring safe working conditions
- making effective use of learning support assistants.

(Association for Science Education, 1997)

The key is to provide a curriculum, with an accompanying pedagogy, that differentiates but does not discriminate.

The Challenge for Science Education

Is there a hopeful future for a science education that takes account of the needs and contributions of all in society? Government documentation provides some signs for hope. For example, although the draft versions of the current Initial Teacher Training National Curriculum for Primary Science were watered down, the final version still states that:

> Trainees must demonstrate that they know and understand the nature of science, including that:
>
> i science is a way of making sense of natural phenomena and as such involves the interaction of an existing body of knowledge with the 'discovery' of new evidence, leading to a re-interpretation of explanation of phenomena and processes;
>
> ii scientific knowledge and explanations may change as new evidence is collected and thinking is challenged.
>
> (DfEE, 1998, p. 78)

Encouragingly, too, the recently published Nuffield Foundation Report on the future of the science curriculum focuses on the need for the Science National Curriculum from age 5 to 16 to be relevant for *all* those taking it (Millar and Osborne, 1998).

Finally, it is always worth being encouraged by the difference that classroom teaching can make to pupils' learning. Many teachers have tried getting pupils to draw what they think a scientist looks like. Lyn Harrison found that when she asked her Year 4 class, 'What is science all about?' two of the conclusions her pupils came up with were:

- best scientists are old men with white coats and moustaches;
- they study a lot and are usually over 60.

(Harrison and Matthews, 1998, p. 22)

In collaboration with Brian Matthews, Lyn Harrison began a study with her class to find out whether such stereotypes could be countered. The children were given worksheets on:

- Mae Jemison – first black female astronaut;
- Louis Latimer – black scientist who assisted in the development of the light bulb and invented improved filaments and bulbs;
- Jocelyn Bell Burnell – astronomer who worked at the Royal Observatory, developed theories on pulsars and quasars, but failed exams at school;
- Elizabeth Garrett Anderson – first woman in Britain to qualify as a doctor;
- Marion North – Victorian botanist who travelled the world and painted plants; her pictures fill a house in Kew Gardens;
- Charles Drew – black scientist who pioneered blood transfusion.

Nearly three months after the teaching, each child was asked to draw two scientists who were working. The children were now more likely to draw female scientists, black scientists and scientists co-operating. It is true that too strong a conclusion should not be drawn from this small study. Indeed, the validity of the 'draw what you think a scientist looks like' technique has been questioned by some on the grounds that pupils may respond with what they believe the interviewer wants them to draw, rather than with what they really think. Nevertheless, the actions of a teacher *can* help children to think in new ways. At St Andrew's school in Cambridge they have a mirror up with the label 'A scientist' attached to it. If you ask pupils what a scientist is like, they say, 'Like me'.

References

Association for Science Education (1981) *Science in Society*, London: Heinemann Educational, and Hatfield: Association for Science Education.

Association for Science Education (1992–3) *SATIS 8–14*, Hatfield: Association for Science Education.

Association for Science Education (1997) *Access to Science Education: Policy*, Hatfield: Association for Science Education.

Baird, V. (1992) 'Difference and defiance', *New Internationalist*, 233, 4–7.

Byers, A., Childs, A. and Lainé, C. (1994) *The Science Teacher's Handbook: Ideas and Activities for Every Classroom*, London: Voluntary Services Overseas, and Oxford: Heinemann Educational.

Byrne, E. M. (1993) *Women and Science: The Snark Syndrome*, London: Falmer Press.

Chalmers, A. (1990) *Science and its Fabrication*, Milton Keynes: Open University Press.

Cunningham, A. and Williams, P. (1993) 'De-centring the "big picture": *The Origins of Modern Science* and the modern origins of science', *British Journal for the History of Science*, 26, 407–32.

Department for Education (DfE) (1995) *Science in the National Curriculum*, London: HMSO.

Department for Education and Employment (DfEE) (1998) *Teaching: High Status, High Standards – Requirements for Courses of Initial Teacher Training (Circular Number 4/98)*, London: Department for Education and Employment.

Donnelly, J., Buchan, A., Jenkins, E., Laws, P. and Welford, G. (1996) *Investigations by Order: Policy, Curriculum and Science Teachers' Work under the Education Reform Act*, Nafferton: Studies in Education.

Driver, R., Leach, J., Millar, R. and Scott, P. (1996) *Young People's Images of Science*, Buckingham: Open University Press.

Edwards, D. and Mercer, N. (1987) *Common Knowledge: The Development of Understanding in the Classroom*, London: Methuen.

Feyerabend, P. (1988) *Against Method*, London: Verso.

Gilligan, C. (1982) *In a Different Voice: Psychological Theory and Women's Development*, Cambridge, Mass.: Harvard University Press.

Goldsworthy, A. (1998) 'Learning to investigate' in R. Sherrington (ed.) *ASE Guide to Primary Science Education*, Cheltenham: Stanley Thornes, and Hatfield: Association for Science Education, pp. 63–70.

Guzzetti, B.J. and Williams, W.O. (1996) 'Gender, text, and discussion: examining intellectual safety in the science classroom', *Journal of Research in Science Teaching*, 33: 5–20.

Harrison, L. and Matthews, B. (1998) 'Are we treating science and scientists fairly?', *Primary Science Review*, 51: 22–5.

Head, J. (1996) 'Gender identity and cognitive style' in P.F. Murphy and C.V. Gipps (eds) *Equity in the Classroom: Towards Effective Pedagogy for Girls and Boys*, London: Falmer Press, and Paris: UNESCO, pp. 59–69.

Henwood, F. (1996) 'WISE choices? Understanding occupational decision-making in a climate of equal opportunities for women in science and technology', *Gender and Education*, 8: 199–214.

Kruse, A.-M. (1996) 'Single-sex settings: pedagogies for girls and boys in Danish schools', in P.F. Murphy and C.V. Gipps (eds) *Equity in the Classroom: Towards Effective Pedagogy for Girls and Boys*, London: Falmer Press, and Paris: UNESCO, pp. 173–91.

Kuhn, T. S. (1970) *The Structure of Scientific Revolutions*, second edition, Chicago: University of Chicago Press.

Mayoh, K. and Knutton, S. (1997) 'Using out-of-school experience in science lessons: reality or rhetoric?', *International Journal of Science Education*, 19: 847–67.

Millar, R. and Osborne, J. (1998) *Beyond 2000: Science Education for the Future*, London: Nuffield Foundation.

Peacock, A. (ed.) (1991) *Science in Primary Schools: The Multicultural Dimension*, Basingstoke: Macmillan Education.

QCA (1999) *The Review of the National Curriculum in England: The Consultation Materials*, London: QCA.

Reiss, M. (1993a) 'Biology-based investigations for AT1', *Offshoots*, Summer: 4–6.

Reiss, M. J. (1993b) *Science Education for a Pluralist Society*, Buckingham: Open University Press.

Reiss, M. J. (1998) 'Science for all' in R. Sherrington (ed.) *ASE Guide to Primary Science Education*, Cheltenham: Stanley Thornes, and Hatfield: Association for Science Education, pp. 34–43.

Siraj-Blatchford, J. (1996) *Learning Technology, Science and Social Justice: An Integrated Approach for 3–13 Year Olds*, Nottingham: Education Now.

Solomon, J. (1983) *Science In a Social CONtext*, Oxford: Basil Blackwell, and Hatfield: Association for Science Education.

Solomon, J. (1993) *Teaching Science, Technology and Society*, Buckingham: Open University Press.

Wolpert, L. (1992) *The Unnatural Nature of Science*, London: Faber and Faber.

11 Possible Futures

Paul Warwick and Rachel Sparks Linfield

In Chapter 2 of this book, Morrison and Webb pointed to the recent tendency to consider science as a 'subsidiary core subject', at least within the primary school. With an increasing emphasis from central government on 'basic skills', and particularly on the development of numeracy and literacy, all other subjects are being placed in the position of having to have the time allocated to them justified by educators. The position of science as a curriculum subject may be seen as being under threat and, ironically, its very success in the UK (Harmon et al., 1997) may well lead some to consider that all is well. In this context, the provision of an exemplar scheme of work (QCA, 1998), tighter subject knowledge requirements within courses of initial teacher training (DfEE, 1998) and the publication of a revised statutory curriculum (QCA, 1999b) may be seen by some as all that is required to maintain the momentum for at least the next decade. This book has demonstrated, however, that the process of curriculum development in science is a dynamic one in which many factors play a part. So let us start by considering government initiatives and their likely effect on the future of science education for the 3–13 age group.

A stress on the importance of early years teaching seems unlikely to diminish, and the structure and content of the early years curriculum will undoubtedly receive periodic review. The replacement of the Desirable Outcomes for Children's Learning (DfE, 1995; SCAA, 1996) with Early Learning Goals (QCA, 1999a) is receiving the critical attention of early years educators throughout the country. Concerns about inappropriate expectations and a flawed structure to the ELGs give substantial pause for thought, yet from a science perspective the changes that are apparent seem primarily to be changes in emphasis. For teachers of very young children, it is impossible to imagine that the interpretation of statutory expectations would not emphasise placing science within real and imaginary contexts (see Coltman's concluding comments in Chapter 3). This is both vital in developing a useful understanding of concepts and fits clearly with the uniquely integrated 'Knowledge and Understanding of the World' section of the Goals (QCA, 1999a). Yet despite such apparently appropriate movements in the responses of teachers to science in the early years curriculum, science is undoubtedly being 'squeezed', at least in the primary phase. Coltman points to the dangers of inadequate time provision within the curriculum for a subject that requires substantial practical work and time to give proper consideration to that work. An erosion of the science curriculum would indeed be ironic, given that the foundations for progress might be seen as stronger than ever before.

Fears about the perceived importance of science within the statutory curriculum as a whole do, however, need to be seen alongside significant attempts by various bodies and individuals to maintain a high profile for science and to move thinking forward with respect to an appropriate science curriculum for the foreseeable future. One major emphasis in the thinking of interested groups has been on a developing acknowledgement of the value of science within society. In Chapter 10, Reiss argues for a science education that 'takes account of the needs and contributions of all in society'. In this context, the development of a scientifically literate population might be seen to be of crucial importance. Such a population would require not only conceptual and procedural understanding, but also the ability to see that understanding as relevant to the contexts provided by day-to-day life and to appreciate the contribution made to that understanding by those of various cultures. The emphasis on inclusion in the new statutory curriculum orders should serve to focus minds on this essential consideration.

In considering the importance of context, Stephenson (Chapter 9) points to the deliberations of those connected to the Nuffield Foundation in their report *Beyond 2000, Science Education for the Future* (Millar and Osborne, 1998). Despite the emphasis within Sc1 on investigations, there has undoubtedly been a move towards a more characteristically traditional 'lower secondary school curriculum' in primary schools, where context and the notion of scientific literacy are sometimes seen as relatively unimportant. The discussion around the publication of *Beyond 2000* revealed an obvious tension between those hoping to promote a 'stories of science' approach within the curriculum, that emphasises above all the development of a scientifically literate population, and the concerns of those who fear a diminution in a 'real science' curriculum that will produce the scientists of the future. This is obviously a simplistic analysis – those wishing to consider the arguments in a little more depth will need to re-read the relevant sections of Chapter 9. Nevertheless, distilling the arguments is very important when it comes to considering the possible 'winners and losers' that arise from the publication of the new statutory orders for science.

This process of distillation is helped by the analysis provided in Chapter 4. Here, Wilson points to the wealth of research and writing that has informed teachers' views of pupils' developing understanding of science concepts. She reflects on how important it is for cross-phase progression for teachers to understand the experiences pupils encounter at each Key Stage and how they move from 'alternative conceptions' (Millar, 1989) to established understandings of science concepts. She makes a case for teaching fewer concepts, so that time can be spent on understanding, rather than on curriculum coverage. In many ways, this reflects Penny Coltman's concerns about the pressurised curriculum. The idea of 'doing a few things well' has considerable appeal when one considers all that pupils need to assimilate in order to *understand* science concepts and procedures. Such a minimalist route, with a focus on context, would be a sensible ultimate goal in a system where pressures from the basics have priority in curriculum intentions for younger pupils and there is continuing discussion about more 'broad-brush', Baccalaureate-style qualifications for pre-university students.

In reviewing the new statutory orders for science, however, it becomes clear that any radical revision along the lines of *Beyond 2000* is a rather forlorn hope. It is undoubtedly true that each Programme of Study places an emphasis on the importance of everyday contexts. Thus, at Key Stages 1 and 2 pupils should be taught 'skills, knowledge and understanding through the use of a range of domestic and environmental contexts that are familiar and of interest to them'. At Key Stage 2 the preamble to the Programme of Study further states that 'they begin to apply their knowledge and understanding of scientific ideas to familiar phenomena, their personal health and things they encounter everyday'. At Key Stage 3 the preamble states that pupils

> understand a range of familiar applications of science. They think about the advantages and drawbacks of scientific and technological developments in environmental and other contexts, considering these developments from other people's perspectives and recognising how opinions may be different in different contexts.
>
> (QCA 1999b, p. 91)

Within the statutory Programme of Study pupils should be taught 'to apply their skills, knowledge and understanding to a range of familiar phenomena, situations and devices' and 'to consider the benefits and drawbacks of scientific and technological developments, including those related to the environment, health and quality of life' (QCA, 1999b).

Such perspectives are undoubtedly important in providing a curriculum that is relevant to learners and will be welcome to many who have increasingly seen work in the science curriculum divorced from any meaningful context. Yet they fall far short of providing a curriculum that is primarily one designed to enhance general scientific literacy, the fundamental objective of those adhering to the philosophy of *Beyond 2000*. This is perhaps not surprising in a context in which many erstwhile proponents of scientific literacy also express fears that a radical revision of the curriculum may affect the quality of future scientists (see Chapter 9), in which there is a clear push to subject specialism at the end of Key Stage 2 and in which end of Key Stage testing, school inspections and exemplar schemes of work (QCA, 1998) are all aligned with an existing curriculum structure.

So what might be seen as the positive aspects of the new statutory orders for science in the National Curriculum? Certainly the inclusions noted above are to be welcomed. They give scope for teachers to continue to place science in real contexts, so important for younger children getting to grips with introductory concepts and procedures. The placing of aspects of the first section of each of the current Programmes of Study ('Sc0' – DfE, 1995) within the body of the revised Sc1, 2, 3 and 4 is a sensible integration of ideas about science and scientific ways of working fully into the statutory framework. Particularly welcome is the fact that all the non-statutory examples cited in the text of the consultation document refer to the use of information and communications technology (ICT) in science. In Chapter 8, McFarlane points to the exciting possibilities that are available to the teacher who embraces new technologies and uses them creatively both as aids to science teaching and also as effective tools through which pupils can become scientists. It is clear that one intention of the revised curriculum is to maximise the use of ICT, and particularly the use of computer technology, within science. This has to be applauded, though we sound a note of warning about budgetary and policy constraints on practice at the end of this chapter.

Considering the teachers of science in 3–13 education, this book highlights the substantial developments that have occurred in recent decades. Research, curriculum development, the influence of external agencies and provision for professional development have all had the effect of moving teachers forward in their understanding of both science and science teaching and learning. Though many writers in this book have pointed to the relative inadequacy of teachers' subject knowledge in science, there can be little doubt that teachers themselves have made Herculean attempts to develop the adequacy of their subject knowledge. This process may well be aided by recent changes to courses of teacher training, which are now required to focus on the adequacy of trainees' subject knowledge as well as their understanding of pedagogy. Yet an understanding of science entails more than a grasp of concepts, and there is a real danger of driving away prospective trainee teachers if the currently proposed mathematics and English test at the end of initial teacher training are extended to embrace science. If pupils are to become effective scientists they need to understand the scientific way of working, and it is in this area that substantial research activity has begun to contribute to teachers' understanding of how to encourage appropriate ways of working. Researchers and trainers (Gott and Duggan, 1995; Goldsworthy and Feasey, 1994) have produced work that makes the reasons for experimentation clear to practitioners, and this is slowly feeding through to both primary and secondary classrooms through the work of bodies such as the Association for Science Education (ASE) and those involved in the initial and in-service training of teachers. A sustained emphasis on the relative importance of Sc1 (DfE, 1995) within the statutory curriculum is necessary if the encouraging signs of progress are to be developed in the future.

However one may respond to recent comments from within Ofsted about the usefulness of educational research (Tooley and Darby, 1998), the focus that the ensuing debate has placed on the potential for teachers to move beyond case study research is important. It seems clear that those most directly involved with the teaching of children have much to contribute to educational research. Teacher/researcher collaborations have pushed forward the boundaries of our understanding about pupils' ideas in science (SPACE reports, various dates) and the continuation of such research would seem to fit well into the current conception of 'useful research' that allows teachers to act on evidence, not hunches (Tooley and Darby, 1998). It is probably not too far-fetched to suggest that if teachers do not become actively involved in classroom research, including both case studies and more wide-ranging enquiries, they are likely to lose their influence on the future direction of the science curriculum. If we take assessment in science as an example, it is clear that the dominant influence in recent years on assessment practice has been the introduction of end of Key Stage statutory assessments. Chapter 6 indicates the importance of formative assessment in science if pupils are to be encouraged to review their ideas critically. It suggests that greater insights might be gained if diagnostic assessment techniques could be developed to help teachers to pinpoint pupils' specific difficulties. Yet Black (1998), referring particularly to the experience in secondary schools, indicates that coverage of the curriculum and the need for pupils to perform well in statutory assessments dominates attitudes to what is seen as important in the classroom. He suggests that teachers' beliefs about learning need to be challenged so that they can see, and argue for, the continuing value of giving pupils feedback that guides their learning in helpful directions. In order for this to happen, not only is it necessary to create the kind of 'curriculum space' argued for in the conclusion to Chapter 4, but it is also vital that teachers are acting as researchers in their own classrooms to build the evidence base for statements made about what is important in science teaching, learning and assessment. As Hobden states in Chapter 7, 'without an understanding of how the child learns and what triggers retention, our teachers can teach facts until they are blue in the face without success'.

In considering the mechanisms for the effective dissemination of such research, Stephenson (Chapter 9) demonstrates how effective external agencies can be for promoting good practice. In particular, professional associations such as the ASE and the SCIcentre, together with a range of special interest groups, can play a powerful role in bringing the practical ramifications of research to teachers in the classroom. Additionally, such groups and organisations have a clear role to play in showing that science is a part of society. If we hope that pupils will be able to make informed judgements about the role of science in society, appreciating the diverse ways in which scientific endeavour impinges on our lives and appreciating the diverse human influences that contribute to this endeavour, there must always be a role for agencies external to the education system in contributing to science education.

A recurring theme throughout this book has been the quality of teacher knowledge in relation to the conceptual and procedural aspects of science. In Chapter 7, Hobden makes a plea for high-quality, regular in-service education for *all* teachers who teach science, not simply those identified as specialists. Yet in an increasingly technological environment what form should this training take? Russell (1997) argues convincingly for teaching in science to move towards greater integration of IT in science learning, with teachers in the position to ensure that multimedia resources 'are used actively, not passively'. We noted earlier that this is a strong perspective in the current National Curriculum statutory orders for science. Yet both policy and budgetary constraints may well prevent such a vision becoming a reality for pupils in the foreseeable future, meaning that we really are in danger, in McFarlane's own words, 'of being overtaken by the more authentic experiences of science available to children outside school'. Teachers will always strive to improve their skill and expertise because they are genuinely concerned to develop pupil learning. It is up to the policy-makers and budget holders to support their efforts if science education is to have a vibrant future in a technological age, and if we are eventually to have a society with both a scientifically literate population *and* large numbers of that population interested in making science their life's work.

References

Black, P. (1998) 'Formative assessment: Raising standards inside the classroom', *School Science Review*, 80 (291): 39–46.

DfE (1995) *Science in the National Curriculum*, London: HMSO.

DfEE (1998) *Teaching: High Status, High Standards: Requirements for Courses of Initial Teacher Training (Circular 4/98)*, London: DfEE.

Goldsworthy, A. and Feasey, R. (1994) *Making Sense of Primary Science Investigations*, Hatfield: Association for Science Education.

Gott, R. and Duggan, S. (1995) *Investigative Work in the Science Curriculum*, Buckingham: Open University Press.

Harmon, M., Smith, T.A., Martin, M.O., Kelly, D.L., Beaton, A.E., Mullis, I.V.S., Gonzalez, E.J. and Orpwood, G. (1997) *Performance Assessment in IEA's Third International Mathematics and Science Study (TIMSS)*, Boston: Center for the Study of Testing, Evaluation and Educational Policy.

Millar, R. (1989) 'Constructive criticisms', *International Journal of Science Education*, 11 (Special Issue): 83–94.

Millar, R. and Osborne, J. (1998) *Beyond 2000, Science Education for the Future*, London: Nuffield Foundation.

QCA (1998) *Science: A Scheme of Work for Key Stages 1 and 2*, London: QCA.

QCA (1999a) 'QCA Consults on Early Years Education', press release, 19 February 1999, also as 'Early Years Consultation Material' – Hyperlink: http://www.qca.org.uk

QCA (1999b) *The Review of the National Curriculum in England: The Consultation Materials*, London: QCA.

Russell, T. (1997) 'The changing nature of knowing and coming to know', *Primary Science Review*, 47: 9–11.

SCAA (1996) *Desirable Outcomes for Children's Learning*, London: SCAA.

SPACE research reports (various dates) e.g. Russell, T. and Watt, D. (1990) *SPACE Research Reports: Growth*, Liverpool: Liverpool University Press.

Tooley, J. and Darby, D. (1998) *Educational Research: A Critique*, London: Ofsted.

Index